Geoengineering

Geoengineering

The Gamble

Gernot Wagner

polity

First published in 2021 by Polity Press

Polity Press
65 Bridge Street
Cambridge CB2 1UR, UK

Polity Press
101 Station Landing
Suite 300
Medford, MA 02155, USA

ISBN-13: 978-1-5095-4305-2 (hardback)
ISBN-13: 978-1-5095-4306-9 (paperback)

A catalogue record for this book is available from the British Library.

Library of Congress Cataloging-in-Publication Data
Names: Wagner, Gernot, author.
Title: Geoengineering : the gamble / Gernot Wagner.
Description: Cambridge, UK ; Medford, MA : Polity Press, 2021. | Includes bibliographical references and index. | Summary: "A bestselling climate economist asks 'is geoengineering worth the gamble to tackle climate change?'"-- Provided by publisher.
Identifiers: LCCN 2021011283 (print) | LCCN 2021011284 (ebook) | ISBN 9781509543052 (hardback) | ISBN 9781509543069 (paperback) | ISBN 9781509543076 (epub)
Subjects: LCSH: Environmental geotechnology. | Climate change mitigation. | Carbon dioxide mitigation. | Pollution--Economic aspects. | Environmental policy.
Classification: LCC TD171.9 .W34 2021 (print) | LCC TD171.9 (ebook) | DDC 628--dc23
LC record available at https://lccn.loc.gov/2021011283
LC ebook record available at https://lccn.loc.gov/2021011284

Typeset in 11 on 13 pt Sabon
by Fakenham Prepress Solutions, Fakenham, Norfolk NR21 8NL
Printed and bound in Great Britain by CPI Group (UK) Ltd, Croydon

For more on the author, visit: gwagner.com

For further information on Polity, visit our website: politybooks.com

Contents

About the author

Gernot Wagner teaches climate economics at NYU, co-authored *Climate Shock*, and writes Bloomberg's Risky Climate column. He was the founding executive director of Harvard's Solar Geoengineering Research Program and served as lead senior economist at Environmental Defense Fund. His writings appear frequently in the *New York Times*, *Wall Street Journal*, *Washington Post*, *Foreign Affairs*, *Foreign Policy*, *The Atlantic*, *TIME*, among many others. Follow his work at gwagner.com

Introduction

Start here – But don't start with geoengineering

The first time I heard about solar geoengineering, I considered the idea nuts. It is. Two decades later – after having worked on the topic at Environmental Defense Fund, helping launch Harvard's Solar Geoengineering Research Program, and doing quite a bit of research and writing on the topic myself – I still think it is a rather healthy attitude to have toward the topic. The entire enterprise seems like a gamble, and a planetary one at that.

Of course, anyone who's been paying attention to what's happening with the rapidly changing climate will recognize that the world is currently playing a different kind of gamble with the planet, and arguably an even larger one.

Geoengineering – in particular, *solar* geoengineering, attempting to cool the planet by sending a small fraction of sunlight back into space, or by increasing the amount of solar radiation that escapes back into space – is no solution to climate change. That much is clear. It does not address the root cause of too much carbon dioxide (CO_2) in the atmosphere, nor the continuing inflow of CO_2 emissions. Geoengineering is a technofix, and a highly imperfect one at that.

Of course, sanitation, too, is a technofix. Without it, cities would not be possible. Modern life is replete with such technofixes. It's often a fine balance between decrying something as a technofix that simply serves to cement the status quo and celebrating an invention as a clear step forward. It is this constant back-and-forth, this constant internal debate, which characterizes many a geoengineering conversation. There is simply no easy answer, no clear line. Even the very idea of working on the topic comes with a number of judgment calls.

A long history of healthy skepticism

All of us having worked on solar geoengineering have stories on how we got to work on the topic. Most came to it hesitantly – some after a lifetime of work on cutting CO_2 emissions.

Geochemist Wally Broecker left an indelible imprint on the climate science community. In 1975, he introduced the term "global warming" into the literature, after the phenomenon had previously been known by the slightly cumbersome moniker "inadvertent climate modification."[1] In a video message, recorded from his hospital bed, for a 2018 "Planetary Management Symposium" at Arizona State University, Broecker said: "If we are going to prevent the planet from warming up another couple of degrees, we're going to have to go to geoengineering." Broecker did not arrive at this conclusion lightly, in what would turn out to be his final address to his scientific colleagues before his passing.

Broecker was, in fact, highly skeptical of solar geoengineering as a possible climate intervention. I remember him having a number of probing questions, when, in 2013, David Keith came to give a talk on the importance of solar geoengineering research at a climate policy seminar at Columbia University's Faculty House. Broecker's main worry, like that of most others, was that mere talk of geoengineering – especially, once again, the "solar" variety – might detract from the need to cut CO_2 in the first place, a concept often called "moral hazard."

It was precisely this worry that had led to a longstanding, self-imposed, unspoken near-moratorium on solar geoengineering research within the scientific community. Broecker had been a key member of the high-powered group that authored a section on CO_2 as part of a 1965 report by President Lyndon B. Johnson's Science Advisory Committee on "Restoring the Quality of Our Environment."[2] The report did not mention cutting CO_2 emissions as a possible option for addressing climate change. Doing so apparently seemed inconceivable at the time. Instead, it mentioned one possible method of addressing the problem: brightening ocean surfaces in an attempt to reflect more sunlight back into space and cool the planet.

In hindsight, this singular focus on solar geoengineering in the 1965 report was a clear mistake, and one the scientific community has overcorrected for over the course of the coming decades. In 1974, Russian scientist Mikhail Budyko first proposed what has since become the most prominent solar geoengineering method: stratospheric aerosols – introducing tiny reflective particles into the upper atmosphere.[3] Budyko's proposal was translated into English in 1977. It was briefly known as "Budyko's blanket," but mentions of it in the scientific literature and especially public climate discourse soon disappeared.

A 1992 National Academies report picks up on the possibility,[4] but it was not until the 2000s that the technology reemerged in broader scientific and climate conversations. After hearing vague mentions of solar geoengineering in the early 2000s, followed by quick dismissals, I first encountered solar geoengineering in earnest shortly after the late Nobel laureate Paul Crutzen wrote his now famous essay presenting stratospheric sulfur injections as a possible way "to resolve a policy dilemma."[5]

The dilemma: Air pollution in the form of sulfur dioxide (SO_2) kills millions each year; it also helps cool the planet. For example, Europe having begun to clean up its air pollution in the 1980s was clearly beneficial. Medieval cathedrals were no longer melting under acid rain. Forests – and people – are healthier. However, the Arctic is now around 0.5°C warmer as a direct result of decreased SO_2 emissions.[6] These are clear tradeoffs.

Crutzen, in his essay, presented this moral quandary. His essay was published jointly with one written by the late Ralph Cicerone, himself a famed atmospheric scientist and then the President of the U.S. National Academies of Sciences, who wrote in support of Crutzen's controversial essay and of further research.[7] While Crutzen and Cicerone's essays did much to lift the self-imposed research moratorium, skepticism throughout the research and policy communities has remained to this day. I would hasten to add that much of that skepticism is, in fact, still healthy. Solar geoengineering is not a topic one should "embrace," in any sense of the term. That goes for policymakers as much as for researchers "merely" trying to answer lingering scientific questions. To this day, much of the skepticism, in turn, can be explained by "moral hazard" worries, a topic we will discuss in depth in Chapter 7.

Narrowing down "geoengineering"

A quick definitional detour is in order here, as "geoengineering" means different things to different people.

In fact, the term is so vague and all-encompassing as to have lost much meaning, despite still being in frequent use. The term "geoengineering" itself is largely an artefact and a result of the term's frequent use in popular discourse. Experts are typically more precise, and for good reason.

Except for the book's cover – *mea culpa!* – I do not use the term "geoengineering" in this book without further explanation, apart from in direct quotations. I instead use either "solar geoengineering" or "carbon removal." The two are sometimes subsumed under the broad heading of "geoengineering," but the two are, in fact, very different. Neither, in turn, is the only term used for either category of interventions.

Solar geoengineering is sometimes also called "solar radiation management" (SRM), "solar radiation modification" (conveniently, also abbreviated as SRM), or traditionally also "albedo modification." It is a largescale, deliberate intervention to cool the planet by sending a small fraction of sunlight back into space, or by increasing the amount of solar radiation that escapes back into space. The plethora of terms here already indicates the problem. While those working on the topic would immediately recognize the abbreviation "SRM," and I have used it myself in peer-reviewed papers and op-eds alike, I will eschew its use here in favor of "solar geoengineering." The reason for this nomenclature is simple: the "solar" modifies the all-too-popular broader term. That doesn't make "SRM" any less accurate. It's just another term for the same idea.

Here it's also useful to dissect the definition a bit further. One operative term is "largescale." Wearing white in the summer does not count, nor does painting roofs or streets white in an attempt to cool cities – though they are all good illustrations of the broader point. Black absorbs heat, white reflects it.[8] Even all of us in any one hemisphere wearing black winter coats or white summer shirts at once, however, does not alter the global climate. Aerosols in the stratosphere do. "Budyko's blanket" – stratospheric aerosols – thus, is the most commonly discussed method, though by far not the only one. (See Part I for more in-depth discussions of different solar geoengineering methods.) More precisely then, I will often refer to stratospheric aerosols as the specific solar geoengineering method.

Sometimes I will also explicitly discuss another set of technologies that are often subsumed under the broader "geoengineering" heading but that are entirely different: a set of techniques typically called carbon removal, carbon dioxide removal (CDR), carbon geoengineering, or direct air capture. All of these technologies remove CO_2 from the atmosphere directly. Their big advantage: they address the root cause of climate change – excess atmospheric CO_2. Solar geoengineering does not. That makes carbon removal an important part of the world's collective climate response, especially given where things stand today. Carbon removal also comes with its own set of important caveats. Many are entirely different from concerns about solar geoengineering. The one area where they do clearly overlap is vis-à-vis moral hazard considerations, their interaction with efforts to cut CO_2 emissions in the first place (see Chapter 7).

One carbon removal technology is planting trees, in turn sometimes subsumed under a broader umbrella of "natural climate solutions." That is surely part of the overall solution, but it can indeed only be one part of it. Planting trees might sound more innocuous than building large industrial facilities to take CO_2 out of the atmosphere; however, it also comes with significant limitations. One of these is the time and space needed to plant the billions of trees needed to make a dent in atmospheric CO_2 concentrations.

Another is permanence. Trees decay, releasing CO_2 in the process. In technical terms, trees help take CO_2 out of the atmosphere, but they keep the carbon in the biosphere instead of returning it to the geosphere. Other carbon removal techniques do, in fact, remove CO_2 from the biosphere entirely.

Meanwhile, even planting trees has now been used as a delaying tactic to avoid doing what's necessary. U.S. Republicans under President Donald Trump, for example, have used their "One Trillion Trees" initiative as a way to detract from the need to cut CO_2 – moral hazard in action, or perhaps better: moral hazard *inaction*. None of this, of course, means that we should not be planting more trees. We should. However, we must not use it as an excuse to delay CO_2 emissions cuts.

A possible role for carbon removal and solar geoengineering

Most importantly, we must stop burning fossil fuels and putting CO_2 into the atmosphere. Nothing else will do. There are indeed other, even more potent, and thus important greenhouse gases. Methane (CH_4), for example, might be more important than CO_2 for the *rate* of global warming – something solar geoengineering, too, has a direct role in affecting (see Chapter 2).[9] Nitrous oxide (N_2O) is similarly more potent than CO_2, around 300 times so on a 100-year timescale. And yes, technically water (H_2O) is the most important greenhouse gas of them all. However, human CO_2 emissions stand alone in their long-term influence on the changing climate.

Cutting CO_2, even to zero, will only stop the further increase in climate impacts. It won't stop them altogether. That immediately leads to another important step: coping with what's already in store. Not unlike both carbon removal and especially solar geoengineering today, mentioning climate adaptation was once considered taboo among many committed environmentalists, and for similar reasons. "Let's stop climate change first," the refrain went in the 1990s, "only then can we start talking about adapting to warming already in store." Even Vice President Al Gore believed as much at the time, considering adaptation a mere distraction. He has long since publicly changed his mind on the topic.[10]

Adaptation, of course, can only go so far. For one, there are the usual endemic inequalities. It's the rich who adapt. The poor suffer. Then there are limits to adaptation. Building a seawall to protect against extreme storm surges is one thing; adapting to one or two meters of sea-level rise by century's end by moving entire cities to higher land within decades is quite another. Parts of Miami are flooding today, on sunny days.[11]

Enter carbon removal, taking excess CO_2 out of the atmosphere and, ideally, putting it back underground, into the geosphere. Carbon removal, meanwhile, comes with important caveats of its own, not least the same kind of moral hazard that beset earlier adaptation conversations. Equally important, much like cutting CO_2 emissions in the first place, removing it from the atmosphere is both slow and, for the most part, relatively expensive.

Solar geoengineering, by contrast, is *fast*, *cheap*, and *imperfect*.[12] These three characteristics make solar geoengineering unique among possible climate policy interventions. They also go to the heart of the solar geoengineering gamble. Little is fully known and, thus, certain. Lots depends on details yet to be worked out, and some may never be known for sure. Governance is key. Each of the three core characteristics figures in this assessment.

Fast, cheap, and imperfect

Fast means that solar geoengineering, fully deployed, could help lower global average temperatures within weeks and months – rather than the years and decades that it would take for CO_2 reductions. For example, Mt. Pinatubo's eruption in June 1992 in the Philippines lowered global average temperatures by around 0.5°C within a year. A year later, temperatures were back to normal and have been rising ever since (see Chapter 2).

Cheap is relative, but most estimates put the direct engineering costs for deploying stratospheric aerosols at a scale somewhere in the single-digit billions of dollars per year. Think of several dozen newly designed planes with large fuselages and enormous wingspans flying missions into the stratosphere around the clock.[13] That's not exactly free, but it might as well be. The direct deployment costs are in the single-digit billions of dollars, compared to cutting CO_2 emissions or removing carbon *ex post*, both typically measured in trillions of dollars. It is cheap enough to ensure that the direct costs do not matter meaningfully in a deployment decision made by the world's governments.

Imperfect is just that: solar geoengineering does not address the root cause of excess CO_2 in the atmosphere. It comes with plenty of potential risks. It might be a really bad idea to contemplate, and worse to actually go through with. Equally important, none of that might matter in light of the first two characteristics, all but pushing the world toward deploying solar geoengineering sooner than most of us might deem possible – or desirable – today.

The combination of *fast* and *cheap* puts solar geoengineering at the exact opposite end of the spectrum from cutting CO_2 emissions in the first place. Whereas cutting CO_2 is all about motivating more people, companies, and countries to do more, solar geoengineering governance is largely about stopping premature deployment – doing it too fast, too much, stupidly.

A gamble worth exploring

One does not need to like solar geoengineering to take the idea seriously. I don't like it. The mere thought of it is scary, as I believe it should be. Somebody somewhere will surely find a way to abuse it. Conceptually, as a foil for ambitious CO_2 cuts, people already have. In 2008, at the height of the most significant U.S. federal climate policy push to that date, Newt Gingrich wrote an op-ed saying how solar geoengineering shows that we don't need to cut CO_2 emissions.[14] If only.

I remember shaking hands with David Keith on Saturday, December 12, 2015 in my living room in Cambridge, MA, agreeing to work on what would turn into Harvard's Solar Geoengineering Research Program. The day is significant for indeed a much more significant reason. It was the same day that the Paris Climate Agreement was gaveled into place across the Atlantic. The irony of the moment was not lost on either of us.

The Paris Agreement has been widely hailed for breathing new life into sluggish global climate negotiations. Nobody thought it would solve climate change. Nothing can, by itself. But the Agreement clearly did show some momentum in the right direction and, after a four-year hiatus here in the United States, the pendulum is once again swinging hard in the right direction, hopefully without avoiding the swing back. All of that momentum toward more ambitious emissions cuts is clearly good, and nothing should take away from it!

While somewhat ironic then, it is precisely against this backdrop of increased global ambition to cut CO_2 emissions in the first place, and a broader understanding of the importance of serious climate action, that solar geoengineering should be discussed.

It must not be either–or. The best approach is a balanced portfolio, where solar geoengineering might have some, at most temporary, role in *mitigating* the worst effects of climate change, while the world cuts CO_2 emissions rapidly – to zero, and then some.[15]

Such a balanced approach may well be wishful thinking. If history – and not just climate history – is any guide, it almost surely is. Fundamental forces hold the world back from doing enough to cut CO_2 emissions. Those same forces push the world to do too much when it comes to solar geoengineering.

Part I

Incentives

1

Not *if*, but *when*

Solar geoengineering turns everything we think we know about climate change and climate policy on its head. For one, there is the link between CO_2 concentrations in the atmosphere and eventual global average temperatures, which itself is highly uncertain. The technical term for this link between concentrations and temperatures is "climate sensitivity." A recent, comprehensive review has advanced our thinking there quite a bit and indeed narrowed the band of uncertainties; alas plenty of uncertainties remain.[1] More on that topic, much more, in my prior book, *Climate Shock*, joint with the late, great Marty Weitzman.[2]

Most importantly for our purposes here, solar geoengineering breaks this link between concentrations of CO_2 in the atmosphere and global average temperatures. It is the only potential climate policy intervention to do so. It also does so highly imperfectly. Solar geoengineering does not tackle the root cause of climate change directly. It does, however, tackle global average temperatures – quickly and cheaply.[3]

That, in a nutshell, is why solar geoengineering is not a question of *if* but *when*. There are few ifs and buts about it.

From "Free Rider" to "Free Driver"

Economics 101 is clear about the cause of excess CO_2 emissions in the atmosphere: the benefits of emitting CO_2 are privatized, while the costs of one's pollution are largely socialized. The solution is self-evident: price CO_2 at the difference between the marginal private and social cost. Arthur Pigou suggested as much in 1920, in his case for rabbits overrunning a communal meadow.[4] The diagnosis is the same.

The term for this Economics 101 principle: the *free-rider* effect. It is in nobody's immediate self-interest to go first and bear the costs of mitigating CO_2. That goes for individuals and companies as much as it does for countries. Why commit to something if others won't?

Economists arguably make too much of a deal out of this one element of the analysis. Political Economics 101 immediately points to vast vested interests as the true hurdle for action. Even if politicians in one country are citing other countries' lackadaisical climate policies as a reason for their own inaction, it typically comes down to domestic politics. In short, the *free-rider* effect may be overplayed. It clearly isn't the full explanation of what is preventing steeper CO_2 cuts.[5] But it surely is one part of the fuller picture.

Much as the *free-rider* effect implies too much CO_2 pollution, solar geoengineering is governed by the opposite fundamental forces. It's not about motivating to act, it's about stopping too much action. Call it the "*free-driver*" effect. Marty Weitzman and I coined the term in a *Foreign Policy* essay memorably titled "Playing God." Weitzman later formalized the idea in a peer-reviewed economic paper.[6] We were by far the first to recognize this fundamental property and to consider it important. As is so often the case with game-theoretic ideas, the first mention goes back to Nobel laureate Tom Schelling.[7] Whatever its name, the fact that solar geoengineering is such a potentially powerful tool relative to its costs makes it a force to be reckoned with.

"Free" is relative

"Free," of course, is a slight exaggeration. Deploying solar geoengineering does come with costs. There are potentially large risks, unknowns, and unknowables.[8]

There are also costs for monitoring and guiding any deliberate, largescale solar geoengineering deployment program. The cost in both money and time is potentially large. That, too, is important – and ought to be a crucial part of any sensible solar geoengineering deployment scenario. Chapter 4 will attempt to paint such a scenario.

Here, I'm simply referring to raw deployment costs – the narrow engineering costs of actually doing the solar geoengineering. Those costs are what the *free-driver* effect captures, and they are indeed cheap – too cheap. But solar geoengineering is not free.

In fact, some of the best estimates put the costs of stratospheric aerosols in the single-digit billions of dollars per year during the early stages of deployment. That's not nothing. It isn't tens, or hundreds, of billions of dollars per year either. In short, done "efficiently," deploying solar geoengineering at scale is within the purview of dozens of countries. The military budgets alone of around 35 countries are at least $5 billion, and 24 have budgets greater than $10 billion.[9] Those estimates entail designing an entirely new plane capable of flying missions – sorties, in aerospace speak – to at least around 20 kilometers up and somewhere within plus or minus 30° latitude around the equator. The origin behind this number is instructive by itself.

Common lore has always been that stratospheric aerosols would be cheap, and that deploying them could be done easily. In fact, word in the (small) solar geoengineering research community was that it could be as simple as modifying a dozen or so existing jets. High-flying business jets could do the trick, invoking images of the crazed billionaire business owner taking the seats out of his Gulfstream – and *voila*.

The origin of this belief is a bit murky, but among the first to explore the topic in earnest was a study conducted by Aurora Flight Sciences, funded by David Keith with money from the Fund for Innovative Climate and Energy Research (FICER), which, in turn, had been provided by Bill Gates. (More on all this later, in Chapter 3.) The resulting report presented calculations for a New High Altitude Aircraft and also concluded that it might be as easy as modifying existing aircraft.[10]

Cue a couple of emails from one Wake Smith, sent out of the blue to David Keith and me around 2016, when we were in the early stages of developing what would turn into Harvard's Solar Geoengineering Research Program. (More on that later, too, primarily in Chapter 3.) Wake introduced himself as having held, among many other accomplishments, the position of former President and Chairman of Pemco World Air Services, a leading airplane modification company. He clearly had a lot of expertise on the subject, he cared about climate change, and he wanted to try to be helpful. We met.

I began our first meeting in the way I tended to whenever I spoke to anyone with any kind of business or finance background: Ours was a research effort; commercial interest would be dangerous. And in any case, there was no commercial case here: "Have you heard of the *free-driver* effect?" Wake assured me he had no financial interest, but that he was, in fact, curious about the free-driver effect. He had read David Keith's book and about how modified business jets could work. From David, verbatim:

> Injection of sulfates might be accomplished using Gulfstream business jets retrofitted with off-the-shelf low-bypass jet engines to allow them to fly at altitudes over sixty thousand feet along with the hardware required to generate and disperse the sulfuric acid.[11]

Wake was skeptical. He didn't want to say so directly, at our first meeting, but he clearly thought such a retrofit wouldn't work. Or rather, that a more powerful engine implied a new plane, a new certification process, the works. For someone who used to run a company modifying planes, this seemed like a different exercise altogether: designing a new plane.

Wake set out to demonstrate that his initial reaction was correct. He spoke to engineers at Airbus, Atlas Air, Boeing, Bombardier, GE Engines, Gulfstream, Lockheed Martin, NASA, Near Space Corporation, Northrop Grumman, Rolls Royce Engines, Scaled Composites, The Spaceship Company, and Virgin Orbit.[12] He did what someone with a deep business background would do: he created a development plan for how one might approach a venture that could design such a plane, finance the development, and see things through from conceptualization to deployment.

We ended up co-authoring a paper describing the process, laying out "Stratospheric aerosol injection tactics and costs in the first 15 years of deployment."[13] The gist was: Existing planes are inadequate. It would take a newly designed plane with a large fuselage and sizable wingspan to transport the material and fly into the lower stratosphere. Moving such a plane from concept to deployment would take the better part of a decade.

None of that is *free*. It would cost billions. But nobody we spoke to had any doubts that it would be possible to do. And the cost figures confirmed the broader sentiment: single-digit billions of dollars per year are, in fact, cheap. Very cheap.

The direct comparison with cutting CO_2 emissions is a problem for many reasons. Timescales is one. While solar geoengineering could lower global average temperatures within months, addressing the root cause by cutting CO_2 emissions and pollution would show effects only over decades and centuries. But it is clear that, while far from *free*, solar geoengineering is indeed very cheap by comparison. The absolute lowest estimates of decarbonizing the world economy come in at around $50–100 trillion.[14] That's the total estimate, not the annual cost, but it is still at least 100 to 1,000 times more expensive than the cost estimates for solar geoengineering. If anything then, solar geoengineering is *too* cheap.

In a rational world, there would be no such thing as *too* cheap. Even if something were indeed free, we would not have to do it if we did not want to. Of course, we don't live in a rational world. To begin with, it's highly unclear who the "we" here is. Who makes the decision? Who might be motivated to pay for such a venture? Equally important: If solar geoengineering is so cheap, and the *free driver* is so dominant, why isn't it happening already?

Sand in the free driver's gears

The fact that even a full-scale deployment of stratospheric aerosols seems incredibly cheap goes hand-in-hand with some incredible economics.[15] "Not *if*, but *when*" is the logical conclusion. But there's one more step worth discussing. If it is indeed so cheap and easy, why hasn't it happened already?

That's akin to the joke about two Chicago economists walking down the street and spotting a $20 bill on the ground. Turns one to the other: "Hey, why aren't you picking it up?" Says the other: "It can't be real. If it were, somebody else would have picked it up already."

Of course, there are plenty of reasons why markets aren't – can't be – as efficient as the simplistic, stereotypical "Chicago-style" economics model might suggest. If they were, there wouldn't be a need for the very business schools that are the academic homes of many of these economists. Nor would there be a need for the management consultancies staffed by graduates of said schools. The Swedes these days are handing out Nobel Prizes to "behavioral" economists for good reason. I put "behavioral" in quotes because, in the end, it's just good economics. Making demonstrably false assumptions of perfect "rationality" isn't. Still, it is worth investigating why solar geoengineering hasn't yet been deployed, especially since it is so cheap.

In short, there seems to be plenty of sand in the *free driver*'s gears. The list of possible explanations is long and often very rational. One such explanation is that politicians might fear opposition from deep greens, environmentalists vehemently opposed to the technology. A slightly different flavor of this argument is that pro-solar geoengineering politicians might first want to signal to environmentalists that they are committed to decarbonization. Or, to up the rationality ante even further, politicians might want to pursue solar geoengineering, but they fear that it cannot be effectively governed at the international level – always a good assumption – and, hence, shy away.

All of these explanations are consistent with the apparent conundrum of too little action. They have all appeared in peer-reviewed solar geoengineering literature, my own academic writings included.[16] They might all just be one too rational. It's not as though the *free-driver* hypothesis states that solar geoengineering happens instantaneously and automatically. That's taking the "rational" Chicago-style explanation quite a bit too literally.

No, Mikhail Budyko, when introducing the idea of stratospheric aerosol injection in 1974, should not have led to it right then and there, nor should have the English translation of his book in 1977.[17] Paul Crutzen and Ralph Cicerone might have lifted the self-imposed moratorium among researchers in 2006, leading to an exponential increase in research interest and publications but, over a decade later, direct global research funding on the topic is still at most around $20 million per year.[18] That compares to the U.S. government alone spending over $2 billion on overall climate science research.[19] It is still early days in solar geoengineering research. Uncertainties abound.

It's similarly clear that it would take many years, perhaps decades, to see anything close to a comprehensive deployment program in action – even when somebody somewhere decides to pull the trigger.

Who decides?

That immediately leads to the biggest question of them all: who would that "somebody" be?

It's tempting to look to national governments to be in the driver's seat. It is they, after all, who are, or at least should be, in the lead on climate policy in the first place. We have already established that dozens would have the financial means to pursue a largescale solar geoengineering deployment program.[20] It is governments who ought to balance solar geoengineering with other urgent domestic priorities, primarily cutting CO_2 emissions. It is also they who ought to coordinate solar geoengineering at the international level. That goes for any multilateral, United Nations-led efforts. That also goes for bilateral talks in any number of constellations. NGOs, businesses, and other private actors matter, some more than others. Ultimately, though, it is governments who set policy.

What if solar geoengineering is not set by governments? For one, there are clearly powerful vested interests that have too much influence on national climate policies. The fossil-fuel lobby is one. Both carbon removal and solar geoengineering might be high up on the agenda if the goal is to delay CO_2 cuts as long as possible. That applies to fossil-fuel companies lobbying democratically elected leaders. It would apply even more so in so-called "petrostates," where the national oil company *is* the government. Saudi Arabia comes to mind, with or without the public trading of Saudi Aramco shares. Oil is the source of the Saudi royal family's power, and it clearly wants to maintain that status quo. The same goes for many other countries in the Middle East and well beyond.

Some of the more enlightened oil majors may have set themselves more or less ambitious decarbonization targets. All of them implicitly or explicitly emphasize "net" decarbonization – at the very least implying that direct air capture or other forms of carbon removal will very much be part of their corporate strategy. Moving only a small fraction of the vast marketing dollars traditionally spent on sowing confusion, or worse, on climate action to lobbying for either carbon removal or solar geoengineering ought to have quite a bit of (undue) influence on national policies.

Then there is direct action by nonstate actors. Billionaires have typically topped that list. David Victor coined the term "Greenfinger" for a "self-appointed protector of the planet."[21] The screenplay writes itself. Greenfinger would have a rather conflicted identity. On the one hand, he would act in defiance of James Bond and his government. On the other, he might well see himself as acting on behalf of humanity, out of a desire to fill a void left by governments' reluctance to deploy solar geoengineering.

The trouble with this picture of a billionaire savior? It's not quite that easy. First, there is the raw math. Annual costs somewhere in the single-digit billions of dollars might be cheap for many governments. But even the average billionaire would deplete his or her wealth quite rapidly. Spending perhaps $5 billion consistently over many years might take a $100 billion fortune. That is a rather exclusive club. Bill Gates, among others, has shown interest in solar geoengineering, helping to fund David Keith's work over the years and contributing $4 million of the first $10 million in funding for Harvard's Solar Geoengineering Research Program, formally launched in 2018. This kind of research is indeed wise. It is also far from anything resembling full-scale deployment. Jeff Bezos made news in early 2020 with a $10-billion climate commitment. He would have to give in the order of that amount every year to sustain a deployment program.[22] Although that may well be theoretically possible, it is far from likely.

Much more importantly, any effort to move toward rapid deployment now would be too premature. Some governments might even consider private moves toward deployment by an act of terrorism and meet such attempts by force.[23] And there are lots of ways to outlaw or otherwise prevent private actors from deploying solar geoengineering against a government's wishes. Billionaires tend not to give money to provoke. On the contrary, anyone wanting to push toward deployment despite formal policies and social norms would truly have to be committed to the cause and, even then, it may not be possible.

All of this at least applies to centralized deployment of stratospheric aerosols, for example by newly designed high-flying planes. That might be the most cost-effective lofting technology known today, but it certainly is not the only one. Nobody knows for sure, as none of these methods has been tested, but anything from high-altitude balloons to rail guns might work. What these alternatives have in common is that, at least for balloons, they are less effective and costlier than planes. They are also highly decentralized methods of deployment. That might have rather high appeal to those seeking to go it alone – whether that involves rogue nations or nonstate actors.[24] Chapter 6 explores this scenario in detail.

For now, let me just say that the *who* of solar geoengineering is very much in contention. More importantly, the *who* may not be a single actor, or even a single type of actor. It may also not be a single solar geoengineering method. Cutting CO_2 is not monolithic. Carbon removal is not either. While solar geoengineering's characteristics lend themselves best to one global, centrally coordinated method, the "rational" implementation policy, detailed in Chapter 4, is far from the only scenario, and it might be far from the most likely one.

The geoengineering dilemma

The prisoner's dilemma is famous for boiling down the conundrum of why two perfectly rational individuals – rational, that is, other than having committed the crimes that put them in this situation in the first place – act selfishly and tell on each other, even though cooperating would be better for them as a whole.[25] Each player acts in their self-interest, given the circumstances. Both end up worse off as a result. It's a simple manifestation of the *free-rider* phenomenon governing CO_2 emissions cuts.

Game theory is stock full of many more such dilemmas. Many attempt to capture the world in simple 2×2 matrices involving payoffs for various actions, some more contrived than others. (Game theory, of course, is not alone. See "Trolley Problem.")[26] Despite some very real limitations, these thought experiments are often useful and instructive, explaining much broader points without all the verbiage. Bear with me. We will use the same logic throughout the rest of this chapter to try to understand the broader climate-policy dynamics at work.

The desire to cut CO_2 emissions, or the lack thereof, can be summarized in a simple 2×2 matrix, as shown in Table 1.1.

In the table, **bold letters** with subscripts represent the moves, letters without subscripts represent the outcomes. It would be a bold move to claim that this table represents all of climate policy, but it does boil down some of the most important logic to its bare essentials. H implies high mitigation, L low. The

Table 1.1. Climate mitigation policy as a result of players' preferred moves: A high-mitigation agreement (H) is only possible if both players choose H over low (L) mitigation.[27]

Moves by players 1 \ 2	H_2	L_2
H_1	H	L
L_1	L	L

outcome is simple: Unless both players agree to want H, the outcome will be L.

Technically, Table 1.1 represents a weakest-link negotiation game. It's the simplest possible way to show why getting to H is so difficult: Why should one nation, state, or other jurisdiction do more than the rest, if the rest will just stick to doing L?[28] While the logic encapsulated in this very question demonstrates the collective-action problem at the core of climate policy, it also immediately shows some pathways to try to overcome this situation. Indeed, books have been written on just that subject. Scott Barrett's *Why Cooperate?* is a good place to start.[29]

First, cutting CO_2 emissions may not be as costly as often assumed.[30] Solar photovoltaic costs alone, for example, have famously declined by around 90% in the past decade alone. That might, in fact, be the most important caveat to our game here, and a hopeful one at that. Much of the delay in climate action, after all, may not be because of the lack of international coordination but because of domestic political obstacles.[31] There, too, of course, solar geoengineering might play a role, invoking a type of moral hazard, or its inverse (see Chapter 7), though that's not the type of interaction modeled here.

Another important caveat is that it may indeed be in some countries' self-interest to pursue more ambitious policies around cutting CO_2 in order to persuade others to do the same. That might happen through traditional channels around carbon pricing policies.[32] It might also happen via supply-side channels, for example China wanting to dominate the market for carbon mitigation technologies, thus moving the energy world from one being dominated by "petrostates" to "electrostates."[33]

If we do take the climate mitigation game as a given, however, solar geoengineering might add a particular wrinkle to these discussions. Assume that each country has one additional move: G. That option is both fast and cheap. Yes, it's highly imperfect, too, but the first two properties alone might lead G to sweep the board. Table 1.2 shows the seemingly inevitable outcome.

The *free-driver* effect, in short, doesn't ask *if* solar geoengineering might one day be used. It points to it simply being a question of *when*. Table 1.2 shows the scariest of possible outcomes: solar geoengineering not just being used in addition to ambitious CO_2 emissions cuts, but possibly even instead of them. A little bit of tradeoff between G and H might well be rational and all-but inevitable in its own right. A total substitution surely is neither rational nor inevitable. Then there's a potentially more consequential twist.

Table 1.2. Climate policy with a solar geoengineering (G) option. Without G, low mitigation (L) dominates high mitigation (H). G dominates both.[34]

Moves by players 1 \ 2	**H_2**	**L_2**	**G_2**
H_1	H	L	G
L_1	L	L	G
G_1	G	G	G

What if geoengineering could lead to a more ambitious mitigation agreement?

As a rule, there's little use in introducing game theory, if it doesn't lead to some seemingly counterintuitive results. The 2×2 matrix here might show why climate mitigation action is hard, but that's about it. It doesn't point to any solutions to the dilemma. The 3×3 matrix, with G for solar geoengineering, does the same. It shows how G will dominate, nothing more. There's no more guidance other than to say that everyone should just agree to cut CO_2 considerably – pick H – and get on with it.

Failing to act on cutting CO_2 emissions – picking L – is scary for the planet as a whole. Slithering into solar geoengineering might be scarier still. With that setup, and with a bit more work to understand what's behind Table 1.2, there may well be a way out of this dilemma.

Focusing on H, L, and G alone has lots of limitations. That's for sure. But sticking with that logic for a bit longer, let's try to rank countries' preferences once again. There are those ranking H ≻ L. (Read the squiggly "≻" simply as saying "preferred to." No other magic there.) That implies large climate concern, at least larger than those not ranking H first. It might also imply that, for this particular player, cutting CO_2 is relatively cheap, again at least relatively speaking. Either way, this player would clearly prefer H.

For those ranking H ≻ L, there are now three options for where G could go: first, second, or third. First implies that G dwarfs all else: G ≻ H ≻ L. That ranking would be bad, for at least two reasons. First, solar geoengineering would win, to the exclusion of any other climate policy; clearly a bad outcome. It also makes the game-theoretic model moot. No need for a 3×3 matrix, the outcome is clear: G wins, unless, for example, nations agree on a strict, enforceable moratorium (see Chapter 8).

G going third, H > L > G, is similarly boring. Now even the low-mitigation scenario is preferred to any solar geoengineering use. That's clearly a possible preference ranking. The more fundamentalist elements of the German Green Party come to mind. They might prefer H to L and, thus, abhor anything that appears like a technofix to the much larger, structural problems of the current fossil-fuel economy. I call this position "boring" not because the position itself is. Far from it. It calls for a radical reorganization of society as we know it. But it does now mean G is sidelined in favor of an exclusive focus on cutting CO_2 emissions.

A third possible ranking is H > G > L, one that ranks G second, possibly far behind H but still (reluctantly or not) above L. Even, or perhaps especially, ardent environmentalists might support this ranking in a fit of desperation, given how far unchecked climate change has proceeded.

As in any game-theoretic setting, a good deal now depends on what the other player does. There, too, are three possibilities. We already know that this player ranks L > H. Once again, G can either go first, second, or third.

With G first, we already know what will happen. Ranking G > L > H yields the same outcome as the other player G > H > L. G dominates. Once again, the only way to prevent solar geoengineering in this scenario is to attempt to ban it: a global moratorium of sorts (see Chapter 8).

What, then, if G is ranked second, implying L > G > H for this player. This now quickly gets more complicated, though not prohibitively so. Table 1.3 shows the complete picture.

Table 1.3. Climate outcomes based on each player's complete preferences. Availability of geoengineering (G) could lead to high mitigation agreement (H, in bold), despite one player preferring low mitigation (L) to H.[35]

1\2	H>L>G	H>G>L	L>G>H	L>H>G	G>L>H	G>H>L
H>L>G	H	H	L	L	G	G
H>G>L	H	H	G	H	G	G
L>G>H	L	G	L	L	G	G
L>H>G	L	H	L	L	G	G
G>L>H	G	G	G	G	G	G
G>H>L	G	G	G	G	G	G

In Table 1.3, go to the third row, which has "L > G > H" as player 1's preference. The only outcomes are L and G:

1\2	H>L>G	H>G>L	L>G>H	L>H>G	G>L>H	G>H>L
L>G>H	L	G	L	L	G	G

Which one it is depends entirely on whether player 2 prefers L > G or G > L, regardless of how they rank H.

The big question is why the player preferring L to H might rank L > G > H. If this player ranks L > H strictly because of costs of cutting CO_2 emissions, L > G > H will be a very real possibility. G, after all, is cheap. We are immediately back to the moratorium, assuming the world doesn't want G to win it all. Ban it, and hope to guide climate policy in a productive direction – toward H, that is.

If that player, however, ranks L > H because they do not believe climate change is a problem worth addressing with aggressive action, L > G > H will be less likely. Why risk G if climate change isn't all that bad to begin with?

Now we are in the third scenario: L > H > G. Zoom into the fourth row of Table 1.3 to see where this might lead:

1\2	H>L>G	H>G>L	L>G>H	L>H>G	G>L>H	G>H>L
L>H>G	L	H	L	L	G	G

The most frequent outcomes are still L and G. If the other player ranks G on top, G wins. Not *if*, but *when*. What's striking, then, is when G does *not* win. That seemingly goes counter to the "not *if*, but *when*" logic.

Let's simplify the table a bit more to see this logic. We can drop the two columns where G is ranked first, and compare the first four columns for when L > H > G (row four of Table 1.3) to the ones when L > G > H (row three):

1\2	H>L>G	H>G>L	L>G>H	L>H>G
L>G>H	L	G	L	L
L>H>G	L	H	L	L

If both players rank L on top, L wins. G doesn't add much to this calculus. Let's drop two more columns, to compare players ranking L first to those ranking H first. Now we're left with exactly four cases:

1\2	H>L>G	H>G>L
L>G>H	L	G
L>H>G	L	H

The first column has two cases leading to L as the outcome. That's when the player ranking H > L also ranks G last. The game essentially collapses to the prisoner's dilemma of yore. G doesn't influence the decision. L wins.

Almost there. We're left with two cases.

With G wedged between L and H for both players, G wins. In some sense, the logic here is simply that the two players can't agree on how much CO_2 to cut, so they would rather settle on G than give the other player what they want in terms of CO_2 cuts. That's a disheartening solution. It's also the one that calls for strong solar geoengineering governance. But it's not the only solution.

If G is ranked below H for those preferring L to H, suddenly, H emerges as the winner. That's true, even though one player still ranks L > H. Here the "availability of risky [solar] geoengineering can make an ambitious climate mitigation agreement more likely." That, in fact, is the title of the paper I wrote with then-Ph.D. student Adrien Fabre, arguing just that.[36] The title of that paper is worth restating: it's the mere availability of solar geoengineering that leads to this outcome. Another key word: "risky." In fact, the riskier is solar geoengineering, the more likely is this outcome.

That mere availability helps break the prisoner's dilemma, the *free-rider* problem. It isn't a guarantee. But the mere possibility is worth pointing out: If G ranks just below H for either player, G might indeed help induce H. That's true even though one player still ranks L > H. Assuming G is not just fast and cheap but also highly imperfect – even those ranking L > H still prefer H to G, putting it last – the mere availability of G might prompt otherwise quarreling parties to opt for H.

All of that is true despite our setup that rigged things against H in the first place. Recall how the weakest-link game setup in Table 1.1 all but guaranteed that L would win.

Enter G, and L is no longer a given. The most likely case with G as an option might still be for G to take all: Somebody, somewhere, will opt to use G, and it will dominate the final outcome. All of that seems to put the burden squarely on governments to rein in tendencies to do too much, too soon – in less-than-ideal ways. (Part

III will explore the urgent need for governance in more detail.)

Meanwhile, as long as G is sufficiently risky and uncertain, it might indeed help to induce H. Solar geoengineering, done sensibly, may be a net positive for the planet, or it might not be. We don't yet know enough. The operative terms here are "risky and uncertain." Solar geoengineering is both. There are lots of ways in which things could go off the rails.

2

What could possibly go wrong?

Suppose we could just dim the sun by around 2% with the push of a button. Global average temperatures would lower enough to offset all of the CO_2 that has been pumped into the atmosphere so far, cooling the planet to pre-industrial levels. That alone would be a radical step for many reasons, with some very real risks and uncertainties. Now, if instead we were to turn down the sun by quite a bit more, at least 8%, and do so for a hundred years, the effect would be Snowball Earth: a runaway cooling due to the inability to retain sufficient heat on Earth, as the planet gets covered by snow and ice. Everything freezes to the equator.[1] Life on Earth as we know it ends.

Snowball Earth has happened before, more than once even, during the 200 million year-long stretch of the Neoproterozoic Era, which began around 750 million years ago.

While it might be possible in theory to inject a sufficient quantity of stratospheric aerosols into the lower stratosphere to dim the sun by 8% or more, such efforts would be eminently detectable – and, given what we know about the consequences, the world community would surely stop them. Playing through any such scenarios might be a fun thought experiment for glaciologists. It might also provide useful lessons for more realistic scenarios, much like dialing other models to a hundred can sometimes yield useful insights into what might happen in more realistic scenarios. But Snowball Earth is not a risk of solar geoengineering worth losing much sleep over.

There are real risks aplenty. Perhaps the most prominent list comes from Alan Robock, a climate scientist most famous for studying the climatic impacts of nuclear war and volcanic eruptions. In 2008, he penned an article for the *Bulletin of the Atomic Scientists* titled "20 reasons why [solar] geoengineering may be a bad idea."[2] Let's take them one by one.

1. *Effects on regional climate*

One of the advantages of solar geoengineering in the form of stratospheric aerosols is its global nature. Applied at or near the equator – or, for example, at 15° or 30° north *and* south – they would spread around the globe within weeks, leading to a near-uniform dimming of the sun.[3] Atmospheric winds, aided by the rotation of the Earth, makes this effect impossible to avoid.

There's one way to muck things up. It would be possible to focus on just one hemisphere, injecting aerosols well north or south of the equator, not both. Doing so would be inadvisable, to put it mildly. Regional solar geoengineering could wreak havoc on rainfall patterns and other weather phenomena.[4] In short: don't. Human error (see point 12 below) is one thing. Doing it intentionally would be quite another.

The global nature of stratospheric aerosols is precisely what makes them seem so attractive. Mt. Pinatubo's eruption in June 1991 in the Philippines injected around 20 million tons of SO_2 into the lower stratosphere. The eruption happened close to the equator, almost exactly at 15° north. That's still not ideal, but it led to near-uniform global dimming, decreasing global average temperatures by around 0.5°C the following year. Indeed, Pinatubo is often invoked as a natural analog to explain how solar geoengineering might work. Not a perfect analog, but close.

Pinatubo, at least, is a seemingly more innocuous example than Mount Tambora's larger eruption in April 1815, which led to the year without summer in 1816. That, in turn, forced Mary Wollstonecraft Godwin, aka Mary Shelley, to spend most of her Swiss summer vacation inside, writing what would later turn into *Frankenstein; or: The Modern Prometheus*. One of her fellow vacationers stuck inside near Lake Geneva was John William Polidori. He wrote *The Vampyre*, which would later inspire Bram Stoker to write *Dracula*. If helping birth Frankenstein and Dracula weren't enough, Tampora's eruption would later also be credited with the spread of both cholera and opium.[5]

Pinatubo's distinction, meanwhile, is that it happened much more recently, with much better data to dissect and papers doing so to this date. (The first time *Nature* put a solar geoengineering paper on its cover was in August 2018. Its subject: Pinatubo's effects on global agriculture.)[6]

A lot has been said about Pinatubo. One area of focus has been its effect on regional climates. The research mimics some of the attribution work around good ol' climate change and its impacts on regional weather phenomena. Yes, floods, droughts, storms, and heatwaves have always existed, but modern statistical techniques now allow scientists to estimate how likely any one particular event is with or without climate change. The U.K. Met Office, led by Peter Stott, is particularly good at these exercises, now producing first estimates virtually in real time during major extreme events.[7] The problem with attributing extreme weather events to any one particular volcanic eruption is that major eruptions are so rare. Few are large enough for their plume to reach the stratosphere. That leaves statisticians in a lurch and makes direct attribution so difficult.

Still, there is some evidence of how Pinatubo has changed regional climates. One particular area of concern: reduced precipitation.[8] And it's hard to find precipitation phenomena more important than the African and Asian monsoons. Billions of people are directly affected. No monsoon, no food. It's no wonder, then, that the monsoon is precisely the event that has garnered particular attention among those studying the possible impacts of both unmitigated climate change and solar geoengineering. Robock himself has studied the phenomenon in depth.[9] One verdict, prominently displayed in an abstract: "Arctic SO_2 injection would not just cool the Arctic. Both tropical and Arctic SO_2 injection would disrupt the Asian and African summer monsoons, reducing precipitation to the food supply for billions of people."[10] The very next sentence: "These regional climate anomalies are but one of many reasons that argue against the implementation of this kind of geoengineering."

Add a few pundits seeking arguments against solar geoengineering, and the headlines write themselves. David Keith, in *A Case for Climate Engineering*, zeroes in on "left-leaning pundits such as Arun Gupta [who] argue that a technocratic elite (he cites me by name) threaten the lives of billions for profit: 'Many scientists fear that pumping sulfates into the atmosphere may cause Asia's monsoons to fail, putting more than a billion people at risk of starvation.'"[11] Gupta focuses on supposed "conflicts of interests" – hence Keith's "for profit" reference. And Gupta is far from alone.[12]

It's easy to see how this type of language grates, and Keith spends a lot of time addressing the monsoon question. He begins by making sure not to blame Robock for emphasizing it, as monsoons, of course, are important. But context – and getting things right in the first place – is important, too.

Here it's hard to do better than Keith's response. I present to you an annotated version of his headline sentences. Keith's opening gambit:

> Indeed, used recklessly, geoengineering could threaten billions with starvation. However, Gupta's narrative ignores *all* studies to date (including Robock's) which suggest that the appropriate use of geoengineering could reduce climate risks to Asian agriculture.

Nothing much to add, other than to say that "appropriate use" does a lot of work here. When gone undetected for a long time, *in*appropriate use or simple "human error" (point 12 below) could cause irreparable harm. Yet, this is all separate from the potential of *appropriate* solar geoengineering. If even appropriate use could wreak havoc, the idea would be a clear nonstarter. If it could help – especially, of course, if it could help a lot – it surely is worth investigating further.

> First, most models of climate change show harmful precipitation *increases* in the Asian monsoon region (think floods and mudslides). If geoengineering can slow or stop this increase in precipitation, then it is delivering a benefit, not a harm.

That is *the* key point. Solar geoengineering decreasing precipitation would be bad only if it were applied in isolation. Of course, the reason we are even talking about solar geoengineering in the first place is because of unmitigated climate change. The goal, after all, is to get back to a world before humans started pumping billions of tons of CO_2 into the atmosphere. Considering *appropriate* solar geoengineering does just that, it's hard to quibble with a decrease in precipitation being an overall benefit.

> Second, a reduction in precipitation need not lead to drought. The amount of surface water found in soils or as runoff in rivers depends on the balance between precipitation and evaporation.

Here, "need not" does a lot of work. The question, of course, is whether *appropriate* solar geoengineering, done moderately, modestly, and managed well, brings monsoon conditions closer to pre-industrial levels. Despite the important addition of focusing on "precipitation minus evaporation" rather than precipitation alone, this is where Keith's argument is weakest, and Robock's concern well placed. Keith continues:

> A climate with increased carbon dioxide that has had temperatures held to pre-industrial levels by geoengineering has less precipitation and less evaporation but it may or may not have more drought.

All true, but also all rather concerning. 1,000 – 1,000 does indeed equal 100 – 100. But what if something is slightly off with either subtraction? Suddenly the balance is off, too. If the overall change is large – from 1,000 to 100 – there's plenty of room for error. If the comparison is better thought of as 1,000 – 1,000 versus

990 – 990, the potential for error is much smaller. In any case, little about the climate is quite as simple. The task, per Keith:

> One must do the analysis before making that claim.

Hard to disagree. A statement that goes for both sides.

> Third, critics overlook a related fact. Extreme climate events, such as droughts or floods, depend on the overall strength of the hydrological cycle.

This is the good old debate technique of turning the biggest weakness into a seeming strength: Yes, getting "1,000 – 1,000" to equal either "990 – 990" or "100 – 100" puts a lot of faith into reducing both sides of the equation equally, something that first needs to be further analyzed. But what if "990 – 990" or especially "100 – 100" is much better to begin with?

It is clear that even a small increase in global average temperatures might produce a large increase in extreme events. Think of a bell curve shifting slightly to the right. The ratio of extremes above a certain threshold increases much faster than the average.[13]

The logic is sound. Keith's statement is still rather bold. In essence, Keith here says that the natural strength of the hydrological cycle, before being affected by global warming, was suboptimal to begin with. That might be right. There had been plenty of extreme weather events entirely without climate change kicking things into overdrive. A world with *no global warming + lots of* CO_2 *+ solar geoengineering* might indeed lead to less precipitation and less evaporation than a world with no climate change, none of the excess CO_2, nor any solar geoengineering. That might well be true. It might also be desirable.

Saying so, however, comes with some clear value statements on its own. It is saying that solar geoengineering allows us an extra degree of freedom to choose a *better* world than the one with which we had started. Keith, of course, knows that, too – both the fact that there is not an extra degree of freedom and that a strong value statement is built into this equation.

It's also easy to see, of course, how all of this might make Robock, and not just Robock, highly uncomfortable. I am, too. (And yes, so is Keith, and virtually any other solar geoengineering researcher I know.) All of it is a hard pill to swallow. In any case, climate scientists certainly need to cede the territory to philosophers, ethicists, and the rest of humanity. Technocratic tinkering alone will not suffice. Back to all of that soon. For now, Keith's final point:

> Finally, models run to date suggest that if used appropriately geoengineering could substantially increase food supply by reducing heat-stress during the early growing season in Asia and Africa – an effect that is one of the most important causes of crop loss as the world warms.

Ah, evidence!

And I'm not being glib here. This kind of evidence, of course, is key. These are the same models that have told us that climate change is bad, that humans are the cause, and – in this specific case – that crop yields under unmitigated climate change will decrease significantly. There's no two ways about it. If one trusts the science of climate change – in good part due to this kind of model evidence – it is then hard to say that there is something fundamentally flawed about also concluding that, for example, crop yield in a scenario of appropriate solar geoengineering will increase.

It is worth dissecting this further.

Most fundamental is initial trust in the science.[14] Without it, there's little more to say here. Conversely, there's little to add other than to say that there's not sufficient reason to mistrust the overall scientific edifice. Blind trust is bad. Not every single study is good. And clearly, it would be wrong to hang the future of the planet on any one individual study's conclusions.

Science is messy. It moves in fits and spurts. It sometimes – often – takes some turns that turn out to be wrong. It is not typically a neat linear path toward uncovering the truth and nothing but the truth. Prior studies get updated, improved upon, and proven wrong. On occasion, studies even get retracted altogether, later proven to have been written with the intent to deceive, or worse. But that's the point: science, overall, is self-cleansing.

There is a lot to the notion that science advances by standing on the shoulders of giants. It is equally true that scientists advance by pointing out the errors in others' work. Different scientific disciplines have distinct cultures on how they go about doing that. Some are polite. Some clearly aren't.

In the end, perhaps the most important reason to trust climate science as a whole and the resulting scientific consensus on climate is because nobody has found that one fundamental flaw in this particular edifice that would bring it all crashing down. Doing so would clearly make anyone's scientific career. Yes, there are those small and sometimes not-quite-so-small questions where a fresh pair of eyes connected to a stellar mind, typically topped off with a hefty dose of determination and effort, come up with a new insight, stunning the field's scientific greats. But overall, the scientific consensus on climate is clear.

And yes, there is an entire field of scientific inquiry – Science and Technology Studies, or Science, Technology, and Society (both abbreviated as STS) – that studies the scientific enterprise, its various motivations and vested interests, and how the resulting knowledge, which in turn is never generated in a vacuum, affects or doesn't affect policies and our daily lives.[15] (Long sentences are, in fact, a hallmark of this particular discipline.)

This ends my waxing poetically about science and my chosen profession.

So no, that one final paragraph in Keith's argument above doesn't settle things. If anything, it's an opening gambit, a call for more research. Keith's paragraph cites a single study.[16] That study, in turn, is co-authored by Ken Caldeira at the Carnegie Institution for Science at Stanford, himself a close collaborator of Keith's. (Keith and Caldeira have both been funded by Bill Gates for their geoengineering research and both are advisors to Gates on energy and climate matters. Caldeira has since left his long-standing academic home at the Carnegie Institution of Science and has joined Gates's office fulltime. Yes, science, too, is not about *who* you know; it's about *whom* you know. And no, that ought not diminish trust in science overall. Advancing the state of overall knowledge is still prized above all else, making science as a whole inherently self-correcting.)

Keith's description of the results is instructive, too. Solar geoengineering "*if appropriately used* ... could substantially increase food supply" (emphasis mine). "Appropriately used," once again, is key. Human error could clearly mess with this assumption (see point 12 below). More importantly, we are back to the question of who decides what is appropriate in the first place. What is appropriate will be different for different people. Perhaps those seeking to maximize crop yield in Asia will have a different metric from those seeking the same in Africa. Perhaps maximizing crop yield conflicts with other priorities altogether. And note how the mechanism here is "reducing heat-stress during the early growing season in Asia and Africa." That statement, in turn, is based on a scientific consensus of sorts that heat stress is "one of the most important causes of crop loss as the world warms." True. But how does it compare to other, possibly conflicting, effects? (More on that under point 4, "Effects on plants," where we will dissect the aforementioned *Nature* cover on Pinatubo's effects on global crop yields.)[17]

Who, then, is right, based on what we know to date? Is Robock right in pointing to the potentially adverse effects of solar geoengineering on the monsoon, prompting many a concerned newspaper headline? Or is Keith correct in his chain of arguments in response? Although unsatisfying, my first response is both. And I am not saying that because I have worked with Keith directly and respect Robock and his work.

All of this goes well beyond the science itself. It is one thing for this set of questions to play itself out in dueling scientific papers, which attempt to improve upon each other's work, and to leave things at that. It is quite another for this to happen in publications titled "20 reasons why geoengineering may be a bad idea" on the one hand, and Keith's *A Case for Climate Engineering* on the other. The titles alone, of course, speak volumes. So does the mode of publication, clearly aimed at reaching an audience that goes well beyond the peer-reviewed literature.

Leaving mode and method aside, there is one much broader point here – even going well beyond the impact of solar geoengineering on the monsoon or crop yield more broadly – that bears repeating: The only appropriate way to look at the impacts of solar geoengineering is in the full context of where the world's climate is heading already.

Viewed in isolation, (solar) geoengineering looks simply mad. Why develop a technology akin to a global thermostat, if it isn't to try to address a real problem? Viewed with this underlying problem front and center, (solar) geoengineering indeed looks very different. That goes for the motivation of scientists studying the topic in the first place. It also goes for some of their conclusions, including the impacts on regional climates.

Put back into the form of a simple equation, the right comparison is not between a world with no climate change and one with solar geoengineering. The correct comparison is *no global warming* on the one hand and *no climate change + lots of* CO_2 *+ solar geoengineering* on the other. The first two elements, meanwhile, are better thought of as *unmitigated climate change*. We no longer have the choice not to add lots of CO_2. That has already happened.

In fact, even more correct would be to expand the equation further and add the other elements of climate policy to the mix. That means *aggressive* CO_2 *cuts* (mitigation). It also means coping with what's in store (*adaptation*), and it includes *carbon removal*. "Mathematically," we are left with:

> *unmitigated global warming + aggressive* CO_2 *cuts + adaptation + carbon removal + solar geoengineering*,

compared to:

> *no global warming*.

That's a complex balance to get right. Lots of moving parts. In fact, there is lots of disagreement even over the status quo – what *unmitigated global warming* itself entails. Cue epic debates over emissions scenarios and so-called "representative concentration pathways," the Kaya identity dissecting the various elements from economic activity to emissions, and hundreds, if not thousands, of academic papers attempting to make sense of where things might go.[18] Both *carbon removal* and, perhaps especially, *solar geoengineering* can only be viewed holistically as part of this much broader set of future scenarios.

Global averages are difficult enough. Attempting to address regional climate impacts does not make things any easier. The first attempt at a comprehensive survey of both the modeling literature and dedicated modeling to address such regional questions was published in *Nature Climate Change* in 2019.[19] The paper's lead author is Pete Irvine, a post-doctoral fellow in Keith's research group at the time. (For a time, my own Harvard office shared walls with Keith on one side and Irvine and a colleague on the other. I'm in the paper's acknowledgments, as is, for example, Ken Caldeira. Small world.)

The paper both analyzes past climate model runs that include solar geoengineering and presents a specific model run done by three of its co-authors, employing Princeton's Global Fluid Dynamic Lab model. Its distinguishing features: It disaggregates regional climate effects particularly well, in addition to being able to model tropical cyclone strength and frequency.

The paper's title summarizes the findings well: "Halving warming with idealized solar geoengineering moderates key climate hazards." The first word, "halving," implies a particular take on how solar geoengineering enters the climate policy picture. Prior efforts have often taken *unmitigated climate change* as given and then put all the burden on *solar geoengineering* to compare results to *no climate change*. That, in many ways, assumes extreme solar geoengineering efforts, dialing things to 100. Here the authors split things down the middle. Half the work is done by *aggressive* CO_2 *cuts*, the other half by *solar geoengineering*. (That, in many ways, is precisely what a prophetically titled paper, co-authored by Keith and Irvine in 2016, suggested as "a research hypothesis for the next decade.")[20] The paper's own model looks at a scenario that first assumes a doubling of atmospheric CO_2. That's the starting point of *unmitigated climate change*, which itself increases global average temperatures by around 2°C. Then it proceeds to halve the resulting temperature increase with "idealized" *solar geoengineering*.[21]

The "idealized" part itself is a simplification. The model does not, in fact, show precisely what would happen when stratospheric aerosols are injected. Instead, it approximates the effect by directly dimming the sun. That's imperfect. It comes with its own caveats. It's what most models of this sort do – and yes, the results have since been confirmed in a model that relaxes this assumption.[22] The paper then analyzes a massive amount of data and dissects what would happen around the globe, degree by degree, longitude and latitude. One variable of particular interest: global surface temperatures in each region of the globe under the various scenarios. That, of course, is standard.

It's easy to find that solar geoengineering is good at decreasing global average temperatures. That's what it's designed to do. It might be a bit surprising that it does so pixel by pixel, but if global warming turns up temperatures almost everywhere, turning down the sun ought to do the reverse. The study shows something similar to be true for temperature extremes. Yearly maximum temperature also goes down when solar geoengineering is added to a mix that achieves half its reduction through aggressive CO_2 cuts. That might be a bit more surprising, but not all that much. Turn down the sun, and extreme heat goes down, too.

Perhaps the more surprising results are what happens to precipitation minus evaporation. In line with Keith's argument, it's not just precipitation alone that matters; it's both combined. (Even just presenting precipitation alone would be misleading. It might make headlines, but it's the wrong metric.) The final variable analyzed across a host of prior model runs is directly linked to extreme weather phenomena, tropical cyclones and monsoon included: maximum precipitation over the course of five days within a year. That's not a perfect measure. Nothing is on its own. But it comes close to a measure of tropical cyclone strength.[23]

The main result: Along all four dimensions, adding (idealized) solar geoengineering to the policy makes the world look more like one without climate change to begin with. What's more: these results apply almost everywhere. Most places around the world look closer to a world with no climate change when solar geoengineering is added.

Most relevant for regional climate effects is what happens, well, regionally. For example, if all climate models agreed that 90% of the world would be better off with solar geoengineering but then it turned out that the Asian monsoon would consistently be a lot weaker across various model runs, that ought to raise a serious red flag. Not so here.

Yes, it's true that the analysis can only tell for "most" points on the map. The rest, by and large, aren't made worse off. Statistics can't tell. The changes are too small. But the key bit here is that model runs *can't agree* about where these points are. That may not be all that comforting at first. If models can't agree, what use is there in even looking at them?

Well, yes, don't look at a global climate model if you care about temperatures in central London on January 1, 2100. Climate models don't determine that. But if 90% of climate model runs show that 90% of the time, temperatures across the south of England, averaged across the first decade of the next century, are significantly higher than they are today, that surely points to where things will go with unmitigated climate change. When those models then show that already hot southern India will be even more so, with many more extreme heat days, that's a sure indication that unmitigated climate change will be bad.

Along the same lines, model runs with *aggressive* CO_2 *cuts + solar geoengineering* do not – cannot – tell us on their own how things will be on January 1, 2100 in either London or Chennai. But if 90% of them say that both southern England and southern India will be closer to conditions with *no climate change*, that surely ought to increase one's level of comfort that solar geoengineering may well have the potential to do some good. This paper essentially says just that.

On its own, it is clearly not proof certain that solar geoengineering would be "good" in any sense of that word. On its own, no paper is. That's not how science works. It also – in a mighty overuse of a triple-negative – doesn't show that solar geoengineering could not, in fact, have negative effects. None of this says it's good for the monsoon, Asian or African. (It most definitely doesn't say it's bad either.) None of it says that solar geoengineering *would* be a good idea.

Still, this paper should, at the very least, help persuade those chiefly worried about Robock's first of "20 reasons why [solar] geoengineering may be a bad idea" that the "may" – to use another overused phrase – does a lot of work here. Sure, it *may*. Or it *may* not. The latest evidence suggested by what, by any measure, is the most comprehensive such analysis to date, points to the latter.[24]

2. Continued ocean acidification

True. That's it. Little more to add, at least for sulfate aerosols. One of the more significant – and often overlooked – effects of climate change is ocean acidification. As Robock says, "continued acidification threatens the entire oceanic biological chain, from coral reefs right up to humans."[25] All true, too. It's one important reason why *aggressive* CO_2 *cuts* must come first. *Carbon removal*, too, could make a real difference here. Little else will.

This also provides a bit of a personal cautionary tale for me, with a cameo of how Twitter can sometimes even be useful. When I joined David Keith at Harvard to work directly on solar geoengineering research, one of the first conversations I remember having with him was about how it's high time to stamp out some preconceived notions. The word "nonsense" came up more than once in that context – and how even many respected scientists share some of these notions that are just plain wrong, or at least woefully outdated.

One such preconceived notion was that solar geoengineering "does nothing to reduce the build-up of atmospheric CO_2." That line wasn't just somebody's throwaway comment. It featured prominently in a 2015 U.S. National Academy of Science solar geoengineering report.[26] What it showed, of course, isn't just that reports of that stature do get things wrong, but, to Keith especially, it showed a deeper, ingrained bias. In an esteemed committee, nobody seemed to question this foregone conclusion, or pay much attention to it, or perhaps both. Ignorance *and* apathy, never a good combination.

Long story short, we ended up publishing a comment in *Nature Climate Change* titled, "Solar geoengineering reduces atmospheric carbon burden."[27] The title says it all. The text says more: "Solar geoengineering reduces the carbon burden, and therefore ocean acidification." The first part is, in fact, true. The latter, as it turned out later, is not.

The first part, solar geoengineering reducing atmospheric CO_2, goes primarily via one very important pathway: carbon-cycle feedback. If one were to use solar geoengineering to stabilize global average temperatures, even while continuing to pump massive amounts of CO_2 into the atmosphere – and no, neither of these two hypotheticals is a realistic scenario nor a good idea – then solar geoengineering *would* play the role of turning off carbon-cycle feedback. Hotter climates release even more carbon from terrestrial ecosystems into the atmosphere. That process would be stopped.

A second potentially large pathway involves emissions from melting permafrost. That, too, could be halted when keeping temperatures constant, despite massive CO_2 emissions. (The third and last mechanism we analyzed is interesting on its own, though relatively small overall: It turns out that fossil-fuel combustion engines are generally more efficient when temperatures are cooler. Keeping temperatures down also avoids additional CO_2 emissions.)

The emphasis of our analysis was clearly to stamp out the preconceived notion that solar geoengineering does not decrease atmospheric CO_2. The ocean acidification line was a last-minute addition, and it turns out to be wrong. I know this now because when I mentioned our *Nature Climate Change* analysis on Twitter, Kate Ricke, a climate scientist at University of California, San Diego, responded saying so. Solar geoengineering reduces atmospheric CO_2 for two reasons. It increases the uptake of CO_2 by land *and* by oceans. The land uptake decreases ocean acidification, the ocean uptake increases it. On balance, Ricke concluded, "it's mostly a wash."[28]

The lesson: Don't overreach when stamping out one preconceived notion and introduce a wrong statement in the process. Solar geoengineering via stratospheric aerosols does not decrease ocean acidification. Robock is right. We must indeed cut CO_2 emissions, and do so aggressively.

3. Ozone depletion

Another important – and justified – concern. The depletion of stratospheric ozone, resulting in the "ozone hole" especially over the Antarctic, is both a well-known global environmental problem and that rare global environmental governance success story. Concern over the ozone hole is also intimately linked to the history of solar geoengineering, not least because Paul Crutzen, the Nobel laureate who, in 2006, effectively broke the long-standing taboo on solar geoengineering research, won his Nobel Prize in Chemistry in 1995 for his work on stratospheric ozone.

Thanks to the Montreal Protocol, signed in 1987, stratospheric ozone has been on the course to recovery. Argentine sheep are no longer turning blind. Australians are no longer contracting skin cancer as quickly. A true environmental governance success story, global treaty and all – signed by President Reagan's Administration, no less.[29] (One large caveat there: Montreal regulates chlorofluorocarbons (CFCs) and hydrochlorofluorocarbons (HCFCs). Crucial innovations by DuPont and others led to hydrofluorocarbons (HFCs) as an alternative to CFCs and HCFCs. The problem with HFCs:

they are a potent greenhouse gas in their own right. The good news on the latter: Kigali. More specifically, the Kigali Amendment to the Montreal Protocol, signed in 2016, and another environmental governance success story in its own right, is beginning to rein in HFCs as well.)

Sulfur-based stratospheric aerosols reflect some sunlight back. Sulfates are also acidic, which in turn helps deplete stratospheric ozone or at least slow its recovery. That is surely a significant tradeoff. It is also not just something close to Crutzen's heart. Many other atmospheric chemists have worked on ozone chemistry over the years. One of them is Jim Anderson, a Harvard chemist whose work helped provide perhaps the clearest proof of the human footprint in stratospheric ozone destruction. He did so by flying instruments on the civilian version of the U-2 spy plane, the ER-2, directly measuring the composition of the stratosphere.[30]

The Anderson Group at Harvard has since turned into the Anderson–Keith–Keutsch group. Keith is indeed David Keith, who had previously also been a post-doc in Anderson's lab. Keutsch is Frank Keutsch, an atmospheric chemist also now focused on solar geoengineering research. Early conversations between Keith and Keutsch focused on just that ozone problem. Keutsch, a chemist at heart, remembers pulling out the periodic table – figuratively at least – and going through elements that might indeed have similar optical properties as sulfate aerosols, but without the adverse ozone impacts.

The resulting paper: "Stratospheric solar geoengineering without ozone loss."[31] The element: calcium. Or more specifically: calcite, aka limestone. Its large advantage over sulfates: It's a base. The paper's main graph shows how solar geoengineering using calcite aerosols "may cool the planet while simultaneously repairing the ozone layer."

As is so often the case, one paper does not prove the case. Subsequent experiments in a lab, by the same research group, showed some significant constraints and highlighted the importance of linking models with laboratory experiments.[32] There is significant research left to be done for any definitive conclusions. Meanwhile, most stratospheric aerosol geoengineering studies, those at least that specify how it might be done, still focus on sulfate rather than calcite aerosols.

It is very early days in the research. And yes, Robock is right to point to stratospheric ozone depletion as a potential real problem with sulfate aerosols. It is. The ozone hole is an important concern. The world has come together to gradually fix the problem. It's important to understand how solar geoengineering might affect that recovery. That includes understanding how solar geoengineering might turn the problem on its head. Too much ozone – an ozone "super-recovery" – turns out not to be good either.

4. *Effects on plants*

Stratospheric aerosols have an effect that goes beyond simply turning down the sun. They also scatter light, making it more diffuse. Compare light under a leafy canopy with direct sunlight. Some plants prefer the diffuse light and thrive in a forest's understory, while others, perhaps most importantly, the main food crops, corn, rice, soy, and wheat, prefer direct sunlight. Solar geoengineering would not have anything close to this effect. You wouldn't be able to notice the difference with the naked eye. Plants, meanwhile, would notice.

This phenomenon also led to the first solar geoengineering *Nature* cover in August 2018. Jonathan Proctor, a Ph.D. student at Berkeley at the time, and his advisor Solomon Hsiang, spearheaded the effort that produced the paper analyzing global crop yields following the volcanic eruptions of El Chichón in 1982 and Mount Pinatubo in 1991.[33] The SO_2 spewed into the atmosphere from El Chichón and Mount Pinatubo created more diffuse light, which lowered yields. Or it would have, were more diffuse light the only effect.

The volcanic eruptions also lowered global average temperatures, which, in turn, is good for plants. On net, the two effects more or less balanced each other out. Soy saw the largest net increase, with temperature effects winning out. Wheat saw the largest net decrease, with the light diffusion effect edging out temperature effects. Tellingly, the volcanoes' (small) effect on precipitation made almost no difference one way or another. All of that should tell us that the effects on plants may well be an important consideration. It certainly deserves significantly more study. Proctor et al.'s study might show a neutral overall effect, but that isn't the point of why to look at solar geoengineering. It's to improve our lot relative to unmitigated climate change.

5. More acid deposition

Another potentially important effect – assuming the aerosol used for solar geoengineering is indeed an acid. If it's a base, the effect may well reverse, much like for ozone depletion.

But since sulfate aerosols have featured prominently in solar geoengineering research to date, Robock is certainly right to point to acid deposition as a potential problem. He rightly says its effect would be small overall, especially compared to traditional air pollution. In fact, one recent calculation points to a factor of 25 difference: "Surface sulfur emission incurs 25 times the exposure from stratospheric injection."[34]

More important is the study's grand conclusion: "Direct, non-climate effects of sulfate injection produce net health risk reduction." The study's authors: Seb Eastham, at MIT at the time after a two-year post-doc at Harvard, where he was also affiliated with David Keith's group, another co-author. But the Eastham paper is not alone. In fact, an earlier paper, co-authored by Robock himself, showed a similar factor.[35] Yes, ocean acidification is important. But solar geoengineering does not meaningfully contribute to the problem. The much bigger factor, of course, is unmitigated climate change due to too much CO_2 in the atmosphere.

A broader conclusion: All of this shows both the prominence and importance of Robock's list on the one hand, and of Keith's and others' research efforts on the other. Here, as elsewhere, it is pretty clear that both Robock and Keith have taken important cues from each other.

The acid deposition question also invokes, once again, Paul Crutzen's essay about tradeoffs between air pollution and stratospheric aerosol geoengineering. In a rational world, it would be seen that the effect from reduced air pollution would dwarf all the rest, for the better. Alas we do not live in such a rational world.

6. *Effects of cirrus clouds*

Now we're quickly entering the territory of the truly unknown. Modifying cirrus cloud is sometimes considered as another possible solar geoengineering technique.[36] The trouble? The science is so uncertain that we don't yet know for sure which way the effect might go. It depends. Some cirrus clouds keep heat in and warm what's underneath. Some might have the opposite effect. But Robock here isn't concerned with cirrus cloud thinning as a possible solar geoengineering technique. He rightly points out the largely unknown effects stratospheric aerosols might have on cirrus clouds. Unfortunately, science doesn't know, which simply calls for much more research to give any kind of definitive answer – there's an emergent theme here.

7. Whitening of the sky (but nice sunsets)

Past volcanic eruptions did show up in paintings – from the Renaissance to Realism. Not the volcanoes themselves, but the resulting redder skies.[37] Edvard Munch's *The Scream* may have been one result. Stratospheric aerosols would have much less of an effect than Krakatau's eruption did. Still, it's an effect.

Whether it merits Robock's warning that "both the disappearance of blue skies and the appearance of red sunsets could have strong psychological impacts on humanity," I will leave as an open question. The whitening of the skies, of course, is a symbol of something much larger. Climate writer extraordinaire Elizabeth Kolbert's latest book is called *Under a White Sky* for good reason.[38]

8. Less sun for solar power

Indeed. One of many, many tradeoffs. Dimming the sun by 1 or 2% and making light more diffuse would have an impact on solar output. Volcanic eruptions are highly imperfect in that regard, as they introduce a lot of other gunk into the stratosphere that deliberate solar geoengineering would avoid. Still, the numbers are potentially significant. Mt. Pinatubo, for example, reduced total sunlight by 2.5%, while reducing direct sunlight by 21% and increasing diffuse light by 20%.[39] Solar geoengineering's impact would likely be at least an order of magnitude less. It might also affect different types of solar power differently. Concentrated solar power needs direct light. Photovoltaic power could operate well under more diffuse light. The full impact on solar power is largely unknown, though the impact surely matters.

9. *Environmental impacts of implementation*

Human activity has environmental impacts. That much is clear. Every energy source – whether low-carbon or not – involves significant tradeoffs, some more so than others. Solar geoengineering would be no different. Even "merely" flying dozens of planes around the clock requires significant effort.

That said, the direct environmental impact of solar geoengineering would surely be several orders of magnitude smaller than the current impact humanity has on the planet, and it would also be orders of magnitude smaller than accomplishing similar goals of lowering global average temperatures and decreasing radiative forcing by virtually any other means. Renewable energy sources like solar and wind, for example, have large land-use implications. (I exclude crashing economies and returning to our caves from the suite of options. Covid-19 is *not* a guide for climate policy.)[40]

If anything, its low direct impact is surely an advantage of solar geoengineering vis-à-vis almost any other approach to lessening humanity's impact on the climate, much like its low direct costs are an advantage. One direct consequence: not *if*, but *when*.

That, in turn, might be highly undesirable. But to point to the absurdity of calling the environmental impact of implementation a problem, not wanting solar geoengineering to be as cheap as it is would result in it having a *greater* direct environmental impact. You don't like that it's just a few dozen high-flying planes? Force it to be billions of small balloons – biodegradable or not – that cover the world in visible detritus. That, too, may be technically possible and economically feasible, but it surely cannot be the answer.[41]

10. Rapid warming if deployment stops

Humanity's impact on the planet is not just due to increasing temperatures – and sea levels, and ocean acidification, and ... The rate of change matters, too.[42] A lot. Adaptation, of course, matters. But what if many species, say, can only migrate at a particular maximum speed? Any faster leads to extinction.

One of solar geoengineering's defining characteristics is its speed. Any semi-rational implementation would surely favor a gradual phasing in to avoid a shock to the system as temperatures decline rapidly. This is also one reason why volcanoes are not a good analog for actual solar geoengineering deployment. They dump too much material into the stratosphere too quickly.

The inverse is also true. In fact, a defining characteristic of solar geoengineering is its reversibility. If solar geoengineering were to be used to decrease global average temperatures significantly, and then, for whatever reason, it stopped being used suddenly, temperatures would increase rapidly.

Termination shock is typically one of the first things critics of solar geoengineering mention as a defining problem. It is indeed a problem, though not necessarily for the reasons typically mentioned. Deliberate or unintentional disruptions to any deployment program might well lead to major disruptions. Natural disasters, terrorist incidents, or rapidly changing politics usually top the list of underlying causes.[43] All are important. All may indeed be overplayed.

The Dutch have managed to maintain their dikes throughout two World Wars and escape largely unscathed over centuries. Dikes have indeed failed in the past,[44] but if the world wanted to continue flying dozens of planes into the stratosphere around the clock, chances are it would be able to do so.

Stories of how this might happen depend a lot on the imagined deployment scenario (see Part II), but they typically go thus: Imagine someone, somewhere, pulling the trigger prematurely. Enter, say, the (hypothetically) hyper-rational Germans, who have objected to solar geoengineering deployment all along. But now that deployment is happening, and the world is rightfully concerned about termination shock, even – or especially – the rational Germans would surely build a spare fleet of planes, just in case something should happen to the deployment program.[45]

That story may well be one too rational but, in any case, the fear of a deployment program suddenly stopping and nobody picking up the slack isn't the true reason termination shock is worrisome. It is worrisome for an entirely different reason, which invokes the realm of the unknown and unknowable.

What if, after years of a gradual ramp-up in a carefully monitored deployment system, scientists suddenly unearthed a major problem that has heretofore remained undiscovered? The proverbial warning lights go off. Pulling back seems like the most sensible choice, but doing so would trigger the specter of termination shock.

None of that implies that solar geoengineering is bad, but it surely implies yet another constraint on actors. The rate of change ought to be limited so as not to trigger other foreseen and unforeseen effects. All of that applies to a rational world. Enter fickle politics, where the newly discovered problem may be anything but rational. Yet some sort of mass hysteria leads those in charge to believe they must act. It's easy to see how termination shock may indeed be a real problem.

11. There's no going back

There is. It's called "termination shock," and I had thought *that* was the problem.

Well, that was a slightly flippant answer. Robock is right that there's no going back in any one particular year – the twelve-to-eighteen-month period when aerosols stay in the stratosphere before falling out by themselves. But, if anything, one defining characteristic of solar geoengineering is precisely its "nimbleness," the fact that it can be "controlled." Solar geoengineering may well be "the world's largest control problem," as an early paper on the topic puts it.[46] That, of course, comes with its own problems, not least of which is:

12. Human error

Yes, the human error of pumping far too much CO_2 into the atmosphere has gotten us into this mess. The fear of human error is surely one of the principal concerns about solar geoengineering as well. Errors abound, not least those of ignoring unintended consequences (see much more detail under point 20 below). Error also enters with "moral hazard"-style thinking, inappropriately trading off solar geoengineering with CO_2 cuts. In other words:

13. Undermining emissions mitigation

Yes, undermining CO_2 cuts is an important problem. It typically comes under the heading of "moral hazard," a problem deserving of its own discussion (see Chapter 7).

14. Cost

Direct costs are indeed low.[47] This has little to do with "advocates" wishing for low costs. There's little advocates could do to make them even cheaper. Lamenting solar geoengineering's costs is like lamenting the environmental impacts of implementation (point 9). Robock, in fact, laments the costs in the context of taking away money from other worthy efforts. There, too, the answer is simply to say that costs are not the problem. If anything, they are *too* low, given how low costs lead to the "not *if*, but *when*" conclusion.

15 and 16. Commercial and military control of the technology

Either or both would indeed be problems – with an emphasis on "control." The fact that somebody will need to manufacture the high-flying plane, say, is clear, and there will be plenty of roles for businesses, as there are in any supply chain.

That said, no there must not be any commercial or military *control* of solar geoengineering technologies. Robock is rightly spooked about being contacted by "the CIA" about geoengineering.[48] Even if that was an independent contractor conducting a study, just the appearance of military control would be bad news. The existence of the Convention on the Prohibition of Military or Any Other Hostile Use of Environmental Modification Techniques or "ENMOD," in force since 1978, is important here. ENMOD itself does not prohibit solar geoengineering research or even deployment, assuming a global governance regime that goes well beyond an individual country attempting to go it alone, without regard for its impact on others. An important component here is for solar geoengineering not to be – or to be seen as – a military project.

The same goes for commercial control. It's also where there needs to be a clear line between carbon removal technologies on the one hand and solar geoengineering on the other. Carbon removal is currently expensive. Commercial interest can help drive down those costs, as it should. That's not the case for solar geoengineering. Patents would be bad, period.[49]

17. Conflicts with current treaties

Earnest legal opinions differ here. What is clear is that there's currently no airtight treaty outlawing solar geoengineering, including ENMOD.[50] None of this, of course, means that there won't be a treaty – or domestic laws, for that matter – attempting to outlaw stratospheric aerosol injection or other forms of solar geoengineering. The legal developments around solar geoengineering are indeed among the most interesting aspects of current discussions. "Watch this space," as they say.[51]

18 and 19. Control of the thermostat and questions of moral authority

We're entering the final stretch of Robock's list, and with it the truly thorny questions. It's one thing to say that solar geoengineering gives "us" control of the global thermostat. Who is that "us"? Who would control the thermostat? Who should? What does it imply for moral and political authority, informed consent, and the very idea of democracy?

The inverse, of course, holds true: What is the legitimacy of a tiny minority of activists (currently concentrated largely in Germany and the U.S.) to attempt to forbid the examination of possibilities around solar geoengineering, interventions that could potentially save millions of lives and preserve ecosystems globally?

That reversal might go a step too far, but it highlights the difficulty of invoking "moral authority" for one side or the other. Anchoring matters. Why is the starting point one of mistrust vis-à-vis those researching and thinking about solar geoengineering, rather than those adamantly opposed? One response here goes back to "moral hazard," as do so many of the moral questions around solar geoengineering. If – since – the system that got us into this mess in the first place is bad, anything must be evaluated as to how it interacts with that existing system.

Perhaps more fundamentally, how does one even talk about solar geoengineering in a world swamped by mistrust of expertise and, more fundamentally, of science itself? In the end, a tiny minority of research papers on solar geoengineering focus on the actual "engineering." The vast majority of research and researchers are situated squarely in the social sciences and humanities, addressing governance and ethical questions, along with broader societal implications.[52]

As a social scientist myself, I certainly believe that attention on the part of social scientists is more than appropriate; though the pendulum may have swung rather far away from the natural sciences. It's good to put natural scientific research into context, but the research needs to be there in the first place. In the end, there's no simple takeaway. Solar geoengineering raises many difficult questions. Almost all involve significant tradeoffs. Addressing some of the more consequential of them is precisely the goal of the rest of this book.

The broader skepticism around experts and expertise does raise one topic promoted by a small but all-too-vocal group: that of "chemtrails." Contrary to all the available evidence, this conspiracy theory claims that contrails (condensation trails produced by aircraft engine exhaust), aren't frozen water vapor. They instead are chemicals. Opinions differ about where these chemicals have come from, who has put them there, and especially why: Some say for weather control. Some indeed say it's for geoengineering already under way. Others go all the way to suggesting this is an attempt at mind control or has the intent of mass murdering the unsuspecting public. Cue death threats to researchers working in the field, and plenty of other small and not-so-small annoyances towards them.

Sadly, belief in chemtrails, it appears, is no longer such a fringe belief. Social media surely doesn't help. It appears to amplify the conspiracy theory quite significantly. Meanwhile, all I can say is that there are plenty of debates to be had on (solar) geoengineering. I'm under no illusion that one "open letter" would do the trick, but here goes:

To: Chemtrail believers

Re: Chemtrails are not the geoengineering debate we should be having

Solar geoengineering is controversial, and for good reason. It describes a set of technologies that seeks to reflect a small fraction of sunlight back into space to cool the planet. The most prominent such technology involves deliberately injecting tiny reflective particles into the stratosphere.

There's a serious debate worth having, both on the science and technology itself and on the societal and policy implications. Unfortunately, in some corners of society valid concerns over the impacts of solar engineering have been overtaken by a different set of fears – various versions of the so-called chemtrails conspiracy theory. According to that conspiracy, solar geoengineering has been happening at scale for years or even decades.

The conspiracy isn't exactly small. Around 60 percent of all social media discourse on geoengineering is conspiratorial, according to co-authored research I published myself.[53] A representative poll of the U.S. public reveals that 10 percent describe the conspiracy as "completely true," another 20 to 30 percent say it is "somewhat true." Belief in the conspiracy appears across party lines, and it can get rather personal – death threats and all.

Most versions of the conspiracy involve planes criss-crossing the skies, spraying toxins, turning ordinary contrails into "chemtrails." Motivations range from weather modification (and yes, there are serious research efforts on that topic too)[54] to mind control or worse. No surprise, Twitter and other largely anonymous online fora allow this community of conspiracy to flourish – necessitating responses showing that no, NASA does not have a "cloud machine" but is instead testing its rocket boosters.[55]

I have no doubt that some who have stumbled upon the chemtrails conspiracy are earnestly looking for the truth. Much in the same way that some who believe that vaccinations cause autism, despite all evidence to the contrary,[56] are motivated by having a close relative suffer from autism, chemtrail conspirators sometimes appear to be looking to learn why a loved one suffers from a respiratory illness. The real answer, sadly often, is indeed air pollution, which kills some three to six million people a year globally.[57] Decreasing that pollution clearly ought to be a global priority.

It is also clear that some of those peddling the conspiracy do so for mercenary reasons – selling ads on their website, or using it to grow their brand and drive page clicks.

Whatever the motivation, the "evidence" presented in favor of the conspiracy does not add up. Conspirators often argue that all one needs to do is look up. Scientists have.[58] What they see are contrails: trails largely made up of condensed water vapor. It is the same effect that occurs when you breathe out on a cold day. If the air is sufficiently cold and moist, a plane's mere turbulence can cause a contrail to form. Adding exhaust from a jet engine aids the process.

Contrails have been with us since the dawn of aviation. The earliest explanation of the science I could find in the popular press is a March 1943 article in *Popular Science* explaining what was then called "vapor trails."[59]

The number of contrails, of course, has since increased dramatically, in line with the number of planes in the sky. And yes, those planes pollute. Each roundtrip flight from New York to San Francisco emits around one ton of CO_2 per economy-class passenger.[60] Sadly, CO_2 is invisible. Were it a smelly pink goo, the world would have acted much sooner on CO_2 pollution. It hasn't, despite amazing progress in slashing other kinds of air pollution.

In fact, some of the progress made in reining in air pollution, such as the sulfur dioxide (SO_2) coming out of smoke stacks, leads to serious climate tradeoffs. While outdoor air pollution kills, it also – inadvertently – counteracts some of the warming effects of CO_2. Removing all such air pollution, while clearly positive for human health, could indirectly cause a lot of harm, as the planet warms even further. The result is what Nobel Prize-winning chemist Paul Crutzen, in 2006, described as a "Catch-22."[61]

It is also, to me personally, the best moral case for solar geoengineering research in the first place.[62]

This is precisely where the real solar geoengineering debate ought to be focused. What are the potential risks and benefits? Would mere talk of solar geoengineering distract from the need to cut CO_2 emissions?[63] Or would such talk be a clarion call to prompt more action on climate mitigation? Reasonable people can disagree and, ultimately, can come down on different sides of the question of whether solar geoengineering could – or should – play a role in an overall climate policy portfolio.

But these arguments are a far cry from claims that contrails are really "chemtrails," that thousands of commercial planes aren't "merely" emitting massive amounts of CO_2 but, for example, are deliberately spraying alumina. Aluminum oxide, in someone's soil, is presented as "evidence" for chemtrails. It isn't. Aluminum is the third most abundant element in the Earth's crust, and aluminum oxide is its most common form.[64] Other supposed explanations are even odder and wholly unbelievable to scientists who have looked at the topic.[65]

All that, of course, raises the question of why we should trust scientists in the first place. Wouldn't they have an incentive to hide evidence if there were a global "chemtrails" program operating somewhere? Well, no – that's just not how science works. Does any one institution have incentives to keep secrets? Sure. But would individual scientists across the world keep some sort of vast "chemtrails" conspiracy a secret?

Scientists aren't all that good at lots of things. Polite, social interactions might be one. But the one thing they are good at is pointing out why others are wrong, and improving on prior knowledge. Pointing out why the broad scientific consensus that the planet is warming and humans are the cause of it is wrong would clearly make a scientific career. The fact that this hasn't happened leaves me comfortable in trusting the scientists' consensus on climate change. The fact that for decades no scientist has shown that ordinary contrails aren't – just that – makes me similarly confident that there isn't anything to the "chemtrails" conspiracy.

The world faces a serious pollution challenge. That goes for SO_2 killing scores today, and it goes for the impacts of CO_2 both today and in the future. There are some serious tradeoffs between the two. That's the debate to have, and anyone I know who does research on solar geoengineering is happy to have it. It's also the kind of debate that anyone with an earnest interest in the future of our planet should want to participate in.

In the end, of course, there's no arguing with the truly crazy. Indeed, this letter had a "viral" moment of sorts in its own right, when I first wrote it on the pages of *Earther* in 2018.[66] Alas, the chemtrails conspiracy isn't going away. In any case, this is the last time I will address it in this book. Thank you for bearing with me.

Back to Robock's list of "20 reasons why [solar] geoengineering may be a bad idea,"[67] and perhaps the most important objection of them all:

20. Unexpected consequences

There may be no better case *against* solar geoengineering than pointing to all the things we don't – perhaps can't

– know, without it actually happening. Here, too, solar geoengineering at first appears like the polar opposite of traditional climate mitigation policy.

Potentially fat-tailed, runaway climate risk in the form of various tipping points implies that focusing on averages may miss large potential climate impacts.[68] For solar geoengineering, the known risks and oft-unknown and perhaps unknowable uncertainties surely point in the opposite direction. The greater its risks and uncertainties, the less likely it is that it would – and especially *should* – be deployed.

But what if solar geoengineering's risks and uncertainties are, at least in part, correlated with those of unmitigated climate change? That is, what if, when one is worse, the other tends to be worse, too? In other words, none of this is about solar geoengineering *in isolation*. It's about solar geoengineering in the context of steadily and rapidly increasing risks from unmitigated climate change. It's all about "risk–risk" tradeoffs.

More importantly, the real question today isn't about solar geoengineering deployment; it's about *research* into the technology. There *higher* solar geoengineering risks and uncertainties would indeed point to *greater* need for solar geoengineering research.[69] The reason is precisely the underlying correlation of climate and solar geoengineering risks. None of that yet covers the truly unknown and unknowables, but there is reason to believe it might be true for some risks. Why?

Most of what we know about solar geoengineering comes from standard climate models. Those models have some clear limitations vis-à-vis solar geoengineering. Stratospheric aerosols have long been modeled in a highly "idealized" fashion, simply modeled as turning down the sun. There's plenty that science doesn't yet know. What science does know shows that there are good reasons for more research, much as there are good reasons to believe that the simple "idealized" model might be a good first approximation of what to expect.[70]

To see the "risk–risk" tradeoffs more clearly, it's instructive to focus on one of the most significant uncertainties of unmitigated climate change. One such uncertainty is the all-important link between concentrations of atmospheric CO_2 and eventual global average temperature increases, a parameter known as climate sensitivity.[71] If climate sensitivity turned out to be much worse than commonly thought, solar geoengineering would also look riskier at first blush. But the very reasons that cause climate sensitivity to be greater would also cause solar geoengineering to be more effective, and the increases are (nearly) proportional to each other. The added climate risk, thus, should be much larger and drive the main result: deploy more solar geoengineering. For now, that simply means: do the research.

Such reasoning, purely based on risk and uncertainty tradeoffs, may yet turn out to be too clever by half. Robock, after all, isn't primarily concerned with risks that show up in today's models – or those that can easily be modeled. He's rightfully more concerned about the true unknowns and unknowables. "Unknowable" here may not mean impossible to know. It may just be difficult to unearth the truth. Past research experience indeed shows how important those are, and how difficult they are to uncover.

Back, once again, to the solar geoengineering *Nature* cover, detailing Pinatubo's agricultural impacts. Leaving aside the paper's details for now, the simple fact that it came out in 2018 should tell us something. The paper came out a quarter century after Pinatubo's effects had been felt.

For many years, it had been taken at face value that stratospheric aerosols primarily do two things: lower temperatures and scatter light. Both are true. It's also true that lots of plants prefer scattered light. What had apparently escaped scientists focused on solar geoengineering was that corn, rice, soy, and wheat, the main food crops, prefer direct sunlight. Instead of solar geoengineering, approximated by past volcanic activity, increasing crop yields, the 2018 *Nature* paper shows that yields remain constant. It apparently took a team of economists, none of whom have done any prior work on solar geoengineering, to show this. Not to diminish their contribution in the least, but this study shows how many low-hanging fruits there are in solar geoengineering research.

The upshot: We simply don't yet know enough to make any kind of definitive decision about whether solar geoengineering overall might be good or bad. We do know much more research is needed.[72] That research needs to include both natural and social sciences. The mere thought of solar geoengineering invokes something clearly visceral. Scientific facts alone can only do so much to assuage those feelings. In the end, solar geoengineering is unnatural and uncertain; it's a technofix in the purest sense of the term.

3

The drive to research

Solar geoengineering is not a new idea. The 1965 report from Lyndon B. Johnson's Science Advisory Committee may not have gotten the technique right – it talked about brightening oceans instead of skies – but it was remarkable nonetheless. While the chapter was titled "Atmospheric carbon dioxide," the "solution" mentioned was anything but.[1] It didn't mention carbon taxes. It didn't mention cap and trade. (That had yet to be invented.)[2] The report only focused on what it called "albedo modification," taking it more or less as a given that CO_2 itself can't be reduced.

We now know that is not the case. Reduce CO_2 we must – and can. In any semi-rational depiction of climate policy, *aggressive* CO_2 *cuts* come first. This alone cuts the link between economic activity on the one hand and CO_2 emissions on the other. That, however, is only the first of many links in the long causal chain from economic activity to the climate damages affecting it.

The path continues from CO_2 emissions to atmospheric concentrations, from concentrations to temperatures, from temperatures to climate damages to human welfare.[3] *Adaptation* tackles the end of the chain – from temperatures to climate impacts to how it affects human welfare. That doesn't mean it should come "last" in a temporal sense. If anything, we should have started more aggressive adaptation measures a long time ago, much like cutting CO_2 emissions.

Carbon removal, in turn, breaks the second link from emissions to concentrations, and *solar geoengineering* breaks the third, from atmospheric CO_2 concentrations to global average temperatures. All that leads right back to the formula comparing *unmitigated climate change* + *aggressive* CO_2 *cuts* + *adaptation* + *carbon removal* + *solar geoengineering* on the one hand to *no climate change* on the other.[4] That is always a good equation to keep in mind, primarily to put solar geoengineering in its rightful place – at the end of any number of climate policies, certainly not as the first line of defense. Unlike the 1965 report to President Johnson, which discusses solar geoengineering in isolation, solar geoengineering cannot – must not – stand on its own.

The 1965 National Academies report to President Johnson may not have done so intentionally, but now, in hindsight, it certainly seems as though it previewed one of the core characteristics of solar geoengineering: low direct costs and high leverage compared to reducing CO_2 emissions or removing it from the atmosphere. It took until 1996 for economist Tom Schelling to draw attention to these core properties.[5] David Keith wrote about them in a sweeping review in 2000.[6] The immediate conclusion: "not *if*, but *when*."

"Taboo" or hiatus?

The history of solar geoengineering research is often described as one involving a long-standing taboo that was finally broken by Paul Crutzen's and Ralph Cicerone's essays in 2006. It could similarly be seen as one with a lot of early attention – from the 1965 presidential report on solar geoengineering to Mikhail Budyko introducing the idea of stratospheric aerosol injection in the 1970s[7] – followed by a period when most climate scientists preferred to focus on cutting CO_2 emissions. Some of that focus on cutting CO_2 was due to a renewed understanding that doing so was both important and possible. Some was due to self-censorship out of fear that too much talk of solar geoengineering would detract from the need to cut CO_2 emissions; what would later be called "moral hazard" – both later in time and later here in the book (see Chapter 7, "Green moral hazards.")

That doesn't mean there was *no* attention paid to the topic. In 1992, it even found its way into an entire chapter of a National Research Council report, which reviewed a slew of possible options – from balloons to planes, rockets, and also space-based mirrors.[8] The latter had just found its way into the literature as an alternative of sorts to stratospheric aerosols.[9] It would clearly be more expensive. It might also avoid some risks associated with "Earth-bound" alternatives, which involve more pollution in the (upper) atmosphere. In 1992, David Keith and long-time collaborator Hadi Dowlatabadi popularized some of these ideas in what they called "a serious look at geoengineering," then still looking at both carbon removal and solar geoengineering together.[10]

No history of solar geoengineering is complete without one Edward Teller. The "father of the hydrogen bomb" and, for example, author of *The Constructive Uses of Nuclear Explosives* began exploring the technology in the late 1990s – in his own late 80s and early 90s.[11] Together with astrophysicist and weaponeer Lowell

Wood and others at the Lawrence Livermore National Laboratory, Teller began to explore stratospheric aerosol geoengineering in a series of modeling studies. The results were promising – too promising.

When Wood presented some early results at a 1998 meeting at the Aspen Global Change Institute, he was the usual provocateur.[12] He joked about how starting a nuclear war would be the best way to stop global warming. He also presented his calculations around how stratospheric aerosols could wipe out all anthropogenic climate change. The idea was met with skepticism, to put it mildly. Ken Caldeira and David Keith, both in the audience, didn't think it could work. Turning down the sun must be different from reducing atmospheric CO_2 concentrations. Caldeira describes leaving the meeting in disbelief, wanting to prove Wood wrong. He couldn't. Or at least, after adding solar geoengineering to a state-of-the-art climate model, Caldeira couldn't find fundamental faults in Wood's reasoning.

The idea gained some traction among a small group of researchers. Caldeira authored several of the first dozen or so papers. Keith wrote a review article in 2000, sweepingly titled, "Geoengineering the climate: History and prospect."[13] Keith's distinction: He not only focused on climate modeling and the technology itself but also focused on implications for climate policy, broader societal questions, and ethics.

Emerging from the shadows

By 2006, a sufficient number of researchers had entered the field to merit a scientific meeting, organized by NASA. Caldeira remembers inventing the term "Solar Radiation Management" – avoiding the term "Geoengineering" to pass muster with the NASA bureaucracy and half-jokingly making it sound as bureaucratic as possible.[14] To Caldeira's own surprise, the term caught on, leading to the relatively prominent acronym "SRM." (The IPCC has since wanted to rename things, but it felt compelled to stick to "SRM" as the abbreviation, settling on "Solar Radiation Modification" as a result. Either concept is identical to solar geoengineering.)

The NASA meeting included researchers, government scientists, and NGO representatives, a mix that would become a hallmark of solar geoengineering meetings going forward. It also shows how many of the most prominent researchers have been well aware of the broader societal implications of the technology. Without environmentalists having a seat at the table from the very beginning, there would be little chance for solar geoengineering ever to be seen as anything other than a nutty idea conceived by crazed scientists. An early concern was interest from those most opposed to cutting CO_2 in the first place. Keith christened the idea "moral hazard."[15] Technically, that is a misnomer. It is more like mitigation deterrence, but the original term stuck. (See Chapter 7, "Green moral hazards.")

One name that often came up in that context was Lee Lane. He helped co-convene the 2006 NASA meeting. He would also soon become a fellow at the American Enterprise Institute, a conservative think tank skeptical of the aggressive CO_2 emissions cuts that science says are necessary. But what's rather striking through this early history of solar geoengineering research is how little attention the idea received from vested interest. Brian Flannery, a climate researcher at Exxon from 1980 through 2011 might have been the one notable exception. But Flannery engaged little directly.

NGOs were more hands-on. In 2009, the Environmental Defense Fund declared geoengineering to be one of three "emerging issues" to watch and engage in. In early February 2010, it organized a science day for its board of trustees, inviting Keith, Alan Robock, and other prominent natural and social scientists engaged in the field. (I joined EDF in 2008 as an economist, leaving in 2016 as lead senior economist, and I co-authored the first internal documents on geoengineering. I also co-organized the 2010 science day.)

March 2010 saw the Asilomar International Conference on Climate Intervention Technologies, the largest geoengineering conference up to that point in time, with almost 200 participants. The venue was significant: In 1975, Asilomar hosted what the announcement for the geoengineering conference described as "the historic Asilomar Conference on Recombinant DNA Molecules that to this day is recognized as a landmark effort in self-regulation by the scientific community." Nobel laureate Paul Berg, chair of the 1975 conference, was an advisor and honorary chair of the 2010 version.

There were clear similarities. Both the original Asilomar conference and what was billed as "Asilomar 2.0" focused on potentially promising yet highly risky and, thus, controversial new technologies. In both cases, there was a split in the research community, with some focusing on the potential, others primarily on the peril. There were also clear differences. I vividly remember the late, legendary climate scientist and stellar science communicator Steve Schneider beginning his remarks with the words, "Many of us wished we wouldn't be here." This wasn't because he thought the technology didn't merit discussion – quite the opposite. To him and many others voicing similar concerns, solar geoengineering felt like an admission of failure on the CO_2 mitigation front.

It was similarly clear throughout the five-day meetings that "self-regulation" would not do. Any calls for it would simply be premature. The resulting conference statement by the Scientific Organizing Committee was largely a call for more research and a statement of how the most important questions were all around the broader societal and ethical implications of solar geoengineering, going well beyond the technology itself. The set of self-regulatory principles that did make it into the consensus document largely echoed earlier "Oxford principles," presenting a rather "high-level" guidance.[16]

One of the members of the Scientific Organizing Committee was the late Paul Crutzen. He had written the 2006 essay asking whether stratospheric sulfate aerosols could help "resolve a policy dilemma": decreasing tropospheric sulfur pollution would have immediate health benefits, but doing so would also warm the planet.[17] It was his essay, in combination with the increased attention from researchers and NGOs alike that brought a lot of mainstream attention to the topic. Perhaps it was OK to talk about solar geoengineering after all.

The subsequent decade showed an exponential increase in interest and in scientific papers about the topic. Crutzen's permission slip attracted many more natural scientists, especially social scientists and humanities scholars, to the topic. By one count, the twelve years after Crutzen's and Ralph Cicerone's essays saw the publication of over 1,200 peer-reviewed articles, white papers, and other scholarship on the topic. By now, the number is over 1,500.[18]

A first outdoor experiment?

The majority of papers on solar geoengineering are firmly rooted in the social sciences and humanities, going well beyond mere technical aspects. Within the climate science of geoengineering, in turn, the vast majority of papers are focused on modeling studies. Some of those vehemently opposed to solar geoengineering find even these modeling studies objectionable. Their reasoning is typically based on a slippery-slope argument: for now, it's climate models; soon it will be lab studies; next up, outdoor experiments.

Enter SCoPEx.

It's short for Stratospheric Controlled Perturbation Experiment, but its symbolism goes well beyond the actual physical experiment. The intention is to fly an experimental platform attached to a balloon into the lower stratosphere. Once there, the experiment would release a tiny amount of aerosols – somewhere up to two kilograms possibly of calcium carbonate, though that may yet change. (The first "engineering" flight might simply release water vapor.) The resulting plume would be roughly 100 meters in diameter and a kilometer long. The balloon, guided by propellers, would then fly through the plume to test the various ways that aerosols might alter the stratospheric chemistry. All that's a tiny amount of material. The material released, if it were sulfur, would be less than one commercial airliner spews out in one minute of flight. The largest direct environmental impact? The lead weights on the balloon, and I'm only half joking here.

SCoPEx would be what's typically known as a "process" experiment. As the project's description on Harvard atmospheric chemist Frank Keutsch's website says, "SCoPEx is a scientific experiment to advance understanding of stratospheric aerosols that could be relevant to solar geoengineering."[19] That largely means it would generate insights that would be helpful to those modeling stratospheric aerosols.

But SCoPEx is much more than that. To some it is a symbol of scientists gone mad. Ray Pierrehumbert, himself a respected atmospheric physicist, has written articles with titles such as "The trouble with geoengineers 'hacking the planet'" and "Climate hacking is barking mad."[20] The former includes subtitles such as "Harvard crosses the Rubicon," referring to SCoPEx, followed by: "Harvard as Greenfinger?" (While Pierrehumbert's objection to solar geoengineering in general and SCoPEx in particular is clear, he is also willing to engage in debate on the topic, including at a 2019 seminar at Harvard.)[21] Pierrehumbert's writings are but one indication of the role that SCoPEx has played in the wider imagination. Add "Bill Gates" to the mix, and the conspiratorial headlines write themselves.[22]

Those most painfully aware of it all, of course, are the scientists working on the project directly. Case in point: The typical scientific experiment does not publish the intention of pursuing such an experiment in a peer-reviewed paper. That is precisely what happened with SCoPEx. It first gained some prominence after the team proposing it laid out its plans in a 2014 paper published by the U.K. Royal Society.[23] That publication drew some early scrutiny, both from fellow scientists as well as NGOs and others engaged in these discussions.

Having been at Harvard from 2015 onward, albeit (crucially)[24] not directly involved in SCoPEx itself, I can say from first-hand experience that very little of the attention paid to SCoPEx by many others was based on the actual facts on the ground. While the scientists involved were working on research plans looking years into the future, activists were hyping the project as *this* being the year where the "Harvard balloon" would fly. When that year inevitably passed without a balloon flying, tweets reached a fever pitch about how whichever blog post had been written the year prior, "unearthing" or otherwise "decrying" the effort was successful in "squashing" it. Any tweet sent in response, attempting to clear up some confusion, was an "admission." Public statements were a "concession" or "admission."

All that was multiplied by less-than-genuine actors ascribing all sorts of motives to those involved in SCoPEx. Perhaps the most easily debunked was that the scientists involved were only doing it to make a buck: "Harvard" was taking out patents or otherwise keeping the intellectual property to control the world's weather. It should have sufficed to point to the 2014 SCoPEx paper to show how that's not the action of someone who is secretively scheming to take over the world. It didn't. John Dykema, the first author of the 2014 paper and SCoPEx project scientist working closely with both Frank Keutsch and David Keith, wrote a direct reply with Keith titled: "Why we chose not to patent solar geoengineering technologies."[25]

One complicating factor: In addition to his research and teaching role at Harvard, Keith is also the founder of Carbon Engineering, a start-up company based in British Columbia developing "air-to-fuels" technology to capture carbon from air and convert it into liquid fuels. That company does hold patents, as it should. The goal is to make the technology (much) cheaper. While carbon removal has nothing to do with solar geoengineering, it sometimes does come under the summary heading of "geoengineering."[26] To the casual observer, and even to some otherwise rather well-informed ones, it is apparently easy to conflate the two – either inadvertently, or deliberately, to sow confusion.

All that has led to the unusual step of convening an external advisory committee for SCoPEx. The committee formally reports to the Dean of Harvard's School of Engineering and Applied Sciences and

Harvard's Vice Provost for Research.[27] Concurrently, and, in part, prompted by the Harvard group's activities, the National Academies convened a panel with the promising charge of: "Developing a Research Agenda and Research Governance Approaches for Climate Intervention Strategies that Reflect Sunlight to Cool Earth." It has since done just that, issuing a set of "Recommendations for Solar Geoengineering Research and Research Governance."[28] The key bits: A call for an open, transparent, integrated solar geoengineering research program funded to the tune of $100 to 200 million over five years, including small-scale outdoor research where warranted. The report also gives a shout-out to SCoPEx's external advisory committee.

This is also where the story ends, for now. Primarily, it is one of many extraordinary measures taken for an admittedly extraordinary experiment that concerns much more than just SCoPEx itself. None of the details matters quite as much as the fact that SCoPEx is as much a governance experiment as it is a scientific one. Both are essential to avoid slithering into solar geoengineering without having done the hard work.

Too fast and/or too slow?

The drive toward solar geoengineering research has been both too fast *and* maddeningly slow. It's easy to see why researchers active in the field want to go faster or why those skeptical of solar geoengineering may want to go slower – or stop research altogether. Then there are the more nuanced positions, each with sensible justifications of their own.

Even those actively engaged in the field sometimes say that things are moving too fast. The recent exponential increase in geoengineering research has been largely driven by climate modeling studies on the one hand and, even more so, by social science studies and governance analyses on the other.[29] There are plenty of modeling questions worth asking. Many uncertainties remain.[30] But the first-order answers are relatively well known.[31]

There is then considerable danger that additional modeling studies investigating a minute detail that may be in itself an important (even if small) scientific modeling advancement simply adds noise to the conversation. All of that is exacerbated by the unfortunate tendency of university press offices to sensationalize research findings in press releases, counting success by "media hits." Media outlets, in turn, are often all too happy to comply. Instead of pointing fingers at someone else's research here, though – and there are plenty of examples to point to – let me try to dissect my own, and the media reaction to it.

Wake Smith and I had been working on a study calculating the costs of lofting sulfate aerosols into the lower stratosphere.[32] Prior efforts had calculated how low the direct costs of stratospheric aerosol geoengineering are, leading to conclusions about geoengineering not being a question of *if*, but *when* it would happen, without proper governance mechanisms. Those prior studies also had their deficiencies, justifying our own effort. So, while our broad overall conclusion was not fundamentally different from what had come before, we thought there was good reason to publish the results. We submitted our write-up to *Nature Climate Change*, where it was promptly rejected without being sent for review. Not unusual. We immediately sent the paper to *Environmental Research Letters* – not quite *Nature*, but a good alternative journal. One good argument for doing so: the most prominent prior costing study had been published there as well.[33] We wrote a carefully worded letter to the editors arguing how the paper provided a small but important advance, meriting publication. Our editors agreed. After a standard review process, the paper was in the pipeline for publication.

Fast forward to early November 2018. The paper is about to be published. The publisher says that it could happen any day now. Meanwhile, the journal's press office sends an email saying that they believe the paper merits a press release, passing along a draft. I edit it. All seems good. The publisher, based in the U.K., suggests a publication date of November 23, which happens to be Black Friday, the day after the Thanksgiving holiday in the U.S. – certainly not the day to publish something if one wants to draw attention to it. I send an email to a handful of colleagues alerting them to the paper coming out, joking about how nobody will pay any attention to it, given the date.

Exactly one media outlet contacts us about the paper before its release. The day before Thanksgiving, Wake and I speak with the *Guardian*'s environment editor, who ends up writing a sensible story that gets posted the next day.[34] It emphasizes the low direct costs, includes plenty of caveats, and generally does a good job putting solar geoengineering and the study itself in context. So far so good.

The next day, Black Friday, climate *is* trending in the news. The latest U.S. National Climate Assessment gets published. It includes the usual warnings of the dire environmental and economic impacts of unmitigated climate change.[35] Climate, thus, is trending in the news. Enter someone with the job description of "trending news producer" at CNN, who put together a "story" with the title: "Dimming the sun: The answer to global warming?" Twelve hours later, our paper was competing with one of the Kardashian sisters for the number one trending news spot on Twitter, and the headlines only became more sensationalist from there.[36] Dozens of local Fox News affiliates suddenly had their own post, largely regurgitating the CNN story. To this date, our paper is among the most downloaded papers at *Environmental Research Letters*. As of this writing, the count stands at 85,000, much of it from Thanksgiving weekend 2018.

I'm not saying any of this to boast. It is a cautionary tale. It also raises some important question as to the speed – and very value – of research. Do I think there was value in writing this paper for its scientific contribution? Yes. Did the media coverage around it do more harm than good? Absolutely. Would I, on balance, write this same paper again, had I known about the harmful impact of the ensuing media storm? I don't know.

One way to look at this is to say that these media skirmishes happen. They pass. Few remember this particular one. The paper, meanwhile, has helped dispel another long-standing misconception that modified existing planes, perhaps even modified business jets, could do the job. They likely can't. It would take a newly designed, high-flying plane – and no, that is not a prohibitive hurdle, neither technically nor economically. Scientists ought not to be dissuaded from pursuing any particular line of reasoning by how it might sound on Twitter – nor, I might add, should the resulting media notoriety be a motivation behind pursuing a particular analysis.

Another view, however, is much more (self) critical: Solar geoengineering, for better or (perhaps largely) for worse, is a highly visible and easily distorted field of research. There are lots of important questions still to be answered – and that's precisely the point: Research ought to focus on those important questions, not on the low-hanging fruit of an additional line on one's curriculum vitae.

It is relatively easy to run an existing climate model one more time, with a slightly different set of assumptions – or, worse, with a set of assumptions believed to yield sufficiently controversial results to merit publication. It is a lot harder to find fundamental flaws in existing modeling efforts, or to build a research program that fundamentally advances the field. Research, thus, may simply be going too fast for the good of the field as a whole, calling instead for much more coordination across researchers and individual research efforts.

Research, meanwhile, might also be too slow. And I'm not arguing here based on low research funding relative to the possible importance of the field.[37] Research might be going too slow, especially to those broadly skeptical of geoengineering a whole, because there is still a relatively small number of researchers active in the field. That, in turn, can lead to group think.

Without launching into a full-on critique of European and North American dominance of science more broadly, it is easy to see how the lack of geographical representation is a clear problem.[38] The SRM Governance Initiative attempts to change that. Its Developing Country Impacts Modeling Analysis for SRM (DECIMALS) fund provides research grants to teams based in Argentina, Bangladesh, Benin, Indonesia, Iran, and elsewhere. The first analysis of impacts on Africa, conducted by an African research team, was published only in 2020.[39] Then there is the much broader sense that there are still too many unanswered questions to draw any definitive conclusions. None of that simply calls for yet another climate model run. It does call for a significant broadening of the research – and of the researchers – in the field.

Research to what end?

The largest question around solar geoengineering research is: "to what end?" Guided by the principle of "not *if*, but *when*," the answer is clear: to get ready for the all but inevitable. The worst possible outcome would be to slither into eventual deployment without sufficient knowledge to steer, or perhaps stop inappropriate deployment before it is too late. One key element is researching the risks and uncertainties: both narrowing and deepening the list of what could possibly go wrong.

As I argue in the prior chapter, Alan Robock's list is a good starting point, but it is indeed only that. Knowing more about which risks are important and are worth focusing on is crucial. It also means dismissing some risks as less salient or as posing solvable problems. At the same time, it is instructive to think through the possible future deployment scenarios that might guide *how* solar geoengineering might indeed be implemented, and what can be done to steer it toward more sensible, careful, "rational" deployment.

Part II

Scenarios

A warning

What you're about to read is wrong. Or rather, it will likely be proven to be wrong. I don't know who will be sworn in as U.S. president on January 20, 2025. I don't know what will happen in the Philippines in 2035 or in Saudi Arabia in 2045. Nobody else does either. Thus is the nature of thinking through future scenarios. They are necessarily made up. They have an element of fiction, while trying to appear realistic. They can also be rather useful for mapping out possible futures and for orienting one's strategy.

Royal Dutch Shell, the oil and gas conglomerate, is famous for its scenario planning. In 1998, it published an internal document, since unearthed by Dutch journalist Jelmer Mommers, that presents two scenarios: "The New Game" and "People Power."[1] The latter envisions a future in the then-distant year 2020, with a youth-led "direct-action campaign against [fossil-fuel] companies." It's a scenario of fossil-fuel companies going the way of tobacco, something surely scary for any Shell planner at the time, and still scary now. The scenario also has an eerie resemblance to today, where we have a strong youth-led climate movement campaigning actively – and effectively – against fossil fuels and fossil-fuel companies.

Did Shell predict Greta Thunberg before her birth? Of course not. Was it useful to think through such a scenario to be better prepared when just such a youth-led campaign sprang into action? Presumably.

That is also my goal here. I make no claim as to the likelihood of any of these scenarios. I do believe that working through them is instructive – not because either scenario is likely, but because the real world may well reflect elements of one, two, or possibly all three of these scenarios captured in the subsequent chapters.

4

"Rational" climate policy

Monday, January 20, 2025. Inauguration day. A new U.S. president takes the reins at 12 noon. The past four years have been marked by a deep economic recession, following the disaster that was the Donald Trump presidency and its abject failure in addressing the Covid-19 pandemic.

President Joe Biden has turned his presidency into the kind of transitional term he had promised all along. By 2023, after two years of a painful depression and reckoning with what had come before, as well as a full-on focus on returning science and fact-based policymaking to Washington, D.C., the economy came roaring back with a vengeance. Guided by green stimulus spending, trillions of dollars have flown into high-efficiency, low-carbon infrastructure the world over. Governments from Berlin to Beijing have used (partial) state ownership of airlines and energy conglomerates alike to reorient their business models toward addressing systemic risks writ large, climate chief among them.[1]

Now, with Biden's Vice President Kamala Harris being sworn in as President, and a supportive Congress and Supreme Court to boot, the United States appears poised to pursue even more ambitious climate policies. The EU, China, and many others are in hot pursuit, trying to outdo each other with ratcheting up their own commitments. India just announced its first absolute CO_2 emissions goal. A mutual understanding of what's at stake, combined with close global cooperation among key governments, has resulted in a global race to the climate policy top. As a result, global greenhouse-gas emissions have declined by between 1 and 3% per year each year this decade.

Even *that* pace is not nearly enough to lead to a decarbonized economy by anywhere close to midcentury, leading to an increased consensus among top Harris administration officials to pursue an additional two-pronged strategy: rapid research, development, and deployment of carbon removal technologies and research into solar geoengineering.

The U.S. Department of Energy is put in charge of developing a national carbon removal strategy and rapidly deploying promising technologies. Aided by mandates and the prospect of fundamental tax reform to establish a national carbon tax of over $100 per ton of carbon dioxide by 2030, a burgeoning private industry is in hot pursuit of increasingly viable business models.

Exxon is on board. After fundamental campaign finance reform made their prior "business model" of regulatory capture unprofitable, a number of oil majors have begun to embrace their role as "carbon companies" in the all-inclusive sense of the term. So far, for the most part, that has involved digging up fossilized carbon and supplying it in the form of oil or gas to the global economy. The new business model involves taking carbon out of the atmosphere and either supplying it in the form of liquid fuels or outright burying it underground.

The U.S. National Oceanic and Atmospheric Administration, meanwhile, has been given an annual budget of around $100 million to build a concerted research effort around stratospheric aerosols. This plan has been the most controversial part of Harris's climate strategy, to put it mildly. But even many greens, by now, are on board, having seen first-hand how even the most ambitious climate plans are insufficient in removing all climate impacts and, thus, suffering.

What had once been known as the "napkin diagram," portraying a rough sketch of multiple climate responses, has since become broadly accepted as a picture of comprehensive climate policy (see Figure 4.1).

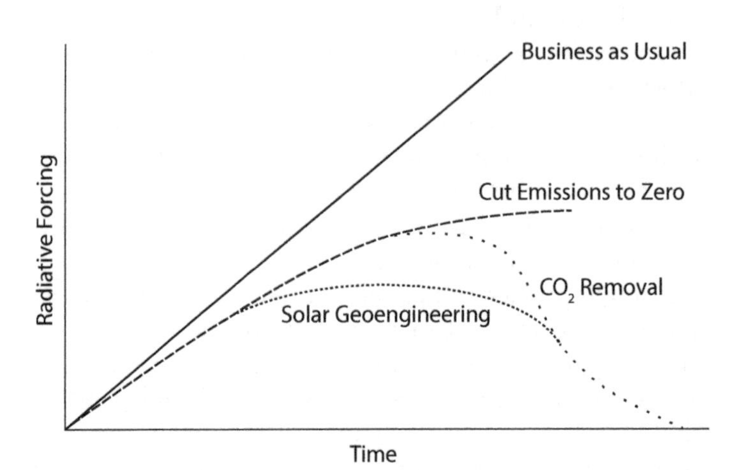

Figure 4.1. Rough sketch of a climate policy portfolio showing climate impacts ("radiative forcing"), including on cutting emissions to zero, carbon removal, and solar geoengineering.[2]

While a comprehensive decarbonization strategy requires significant global cooperation, pursuing anything close to a rational solar geoengineering strategy requires even more. For better and (so far, largely) for worse, the actions of one or two countries – even the United States and China, say – cannot rapidly change the climate of the entire globe. That's what makes mitigating carbon dioxide emissions so difficult. Carbon removal, too, is slow. While there is hope – and need – for rapid technological breakthroughs, there is no such thing as an accidental, sudden over-deployment of carbon removal strategies. Solar geoengineering is in an entirely different category.

Any globally coordinated solar geoengineering effort does indeed require close global coordination among all (major) players involved. That might involve a handful of countries or many more countries and entities such as the European Union as a whole.[3]

One rational approach might be to coordinate any efforts at the highest levels possible. Ideally that might mean the UN General Assembly, with a resolution authorizing a coordinated deployment effort. Any number of other multilateral institutions might be similarly engaged, from UN agencies to other multilateral fora, such as the G20.

It might further be desirable to try to take politics as much as possible out of immediate research oversight. That might involve a global consortium of National Academies of Science or perhaps an entirely independent "blue ribbon" advisory panel. Then again, in our hypothetical, hyper-rational world, that may be neither necessary nor desirable. After all, politics – at least the semi-rational sort – ought to play a role in guiding solar geoengineering more broadly.

A quick reality check

All of this, of course, is fiction. There is no such thing as "rational" climate policy in the real world. There's no single "social planner" all-too-oft assumed in social science and especially economic models. Even if there was, it would be hard to agree on the details of general climate policy and solar geoengineering in particular.

Who sets the thermostat? And to what temperature? The sketch in Figure 4.1 makes a number of important choices along the way, not least the ordering of mitigation first, with carbon removal making up the difference to any agreed-upon climate goal, for example (eventually) limiting global average warming to 2°C. Limiting warming to 1.5°C would imply significantly more ambitious mitigation and carbon removal, or significantly more solar geoengineering, or (likely) both. Even in a perfectly rational world, all these questions would be difficult to answer. In the real world, answering them is nearly impossible.

The real question then is not whether this "rational" climate policy scenario makes sense, but what it would take to approximate a global climate policy akin to the one sketched out in Figure 4.1: Which forces are at work that might push the world toward such a scenario, and which forces would push the world away from it?

One thing is clear: any kind of semi-rational climate policy begins with a significant ramp-up in mitigation. Solar geoengineering cannot make up for the entire, ever-widening gap between a business-as-usual pathway and a stabilized global climate. At best, it could help shave off the peak of climate impacts, serving as a lengthy bridge from the present to a low-carbon future.

In fact, any "rational" deployment program would start relatively slowly, for example by lofting 100 thousand tons of sulfur into the lower stratosphere and dispersing around 200 thousand tons of SO_2. That amount would increase linearly by around 200 thousand tons of SO_2 per year. Given today's rate of temperature change, that is akin to attempting to halve the increase in global average temperatures through this stratospheric aerosol deployment program, with the other half achieved by cutting CO_2 emissions.[4]

Governance is key; what is governance?

All of that clearly demonstrates that the technical and scientific aspects of solar geoengineering, while important, are a mere starting point. Most important questions are centered around governance of whichever form the eventual technology – or set of technologies – might take.

Governance, however, cannot wait for deployment. Technologies don't grow into a vacuum. They are formed by individual research agendas, funding priorities, and institutional influences. All of these also bring research governance and coordination to the fore. It's one thing to sketch a picture of possible climate futures on the back of a napkin. It's another to devise a comprehensive research strategy that might lead there.

The global consortium of National Academies of Science might have been entirely hypothetical, but it may not be far from what is beginning to happen. One of the very first comprehensive geoengineering reports, in 2009, came from the U.K. Royal Society.[5] The first report by the U.S. National Academies of Science that mentioned geoengineering was published in 1992.[6] The most comprehensive National Academies report on solar geoengineering in general came in 2015.[7] The latest, published in 2021, focused on developing a research agenda and governance.[8] Meanwhile, both the Chinese Academy of Sciences and the Chinese Academy of Social Sciences are funding research efforts in China, while the SRM Governance Initiative is doing so mostly in developing countries. None of these efforts on their own are sufficient for a truly global research coordination, but they are important building blocks.

Then there are a number of private fora and initiatives cropping up. Perhaps the most significant of these is the Carnegie Climate Governance (C2G) Initiative, led by Janos Pasztor, formerly a top climate advisor to UN Secretary General Ban Ki-moon and Policy Director at the World Wide Fund for Nature. C2G provides an important catalyst for discussion and international coordination. It is difficult to attribute specific advances to any one group, but it is clear that C2G has played a major role in furthering conversations. At first, its focus had been on both carbon removal and solar geoengineering, introducing both topics into important conversations such as the UN Sustainable Development Goals.[9] By 2020, conversations around carbon removal had become so embedded in climate conversations that C2G now focuses exclusively on solar geoengineering. One important feature of C2G is that its mission is at once broad and still clearly defined, focused on "expanding the conversation from the scientific and research community to the global policy-making arena."[10] That also means that C2G has set an expiration date for itself, in the next 3–4 years. Once its mission of introducing the conversation into various fora, including potentially the UN General Assembly, has been fulfilled, C2G will cease operations.

C2G's mission, in many ways, is precisely what governance is and ought to be: a forum for conversation among those guiding global climate policy. No more, certainly also no less. The field is replete with policy entrepreneurs, who attempt to push particular viewpoints. Good science, too, needs such advocates. Yet ultimately it is up to politicians to make the call, hopefully one guided by sound science and an understanding of shared future for the planet.

5

A humanitarian cyclone crisis

Shift from the "rational" to the real world, and from 2025 to 2035, 2045, or perhaps much closer to today after all.

Manila is recovering from another record tropical cyclone. Millions lost their homes, again. The Filipino economy has yet to recover from last year's record storm season – with three super-cyclones hitting within two months. The Philippines, of course, is not alone. Mozambique has seen a similar increase in cyclone intensity, fueled by unusually high ocean surface temperatures off its coasts. The Gulf of Mexico is in a permanent state of emergency, and there are record temperatures the world over.

Saudi Arabia has cancelled this year's Hajj pilgrimage due to public health concerns. But unlike the last time the Hajj had been cancelled, in 2020 due to concerns over the Covid-19 pandemic, this year's cancellation was solely due to record temperatures. Public health experts are warning that record heat engulfing the Middle East would lead to tens of thousands of heat-stroke deaths if the Hajj were to proceed. India's Independence Day met a similar fate, with New Delhi cancelling festivities and issuing shelter-in-place orders for close to half a billion people – to say nothing of above-average droughts, wildfires, and other weather extremes the world over, all-but cancelling Austria's ski and Australia's beach season all at once.

"Above average," of course, has lost all meaning. Weather extremes are the new normal.[1] While none of this is a surprise, even the most astute climate scientists have been blindsided by the severity of it all. A now-famous 2020 study projected 1 to 3 billion people "to be left outside the climate conditions that have served humanity well over the past 6,000 years" by 2070.[2] That date has kept moving forward, leading to untold suffering along the way.

Even the most aggressive city-level adaptation plans have had to be re-adjusted every few years – and not because of the ever-accelerating rise in sea levels. That alone would be bad enough, and it has city planners scrambling. What's worse are the rapidly increasing extreme events. Versions of the iconic 2007 IPCC graph have since become a mainstay of middle school history books and daily weather reports alike (see Figure 5.1).

It is precisely the frequency of extreme climatic events that has led a handful of countries to consider increasingly desperate measures. Australia has started early Marine Cloud Brightening field tests in an attempt to save the last remnants of its no-longer-all-that-Great Barrier Reef.[4] The effort has shown some promise, but signal has never risen above the noise, making definitive conclusions difficult.

Meanwhile, talk has shifted to early action on another promising yet risky and still-untested method: stratospheric aerosols. The Philippines has expressed interest in serving as a launch site. Australia has pledged

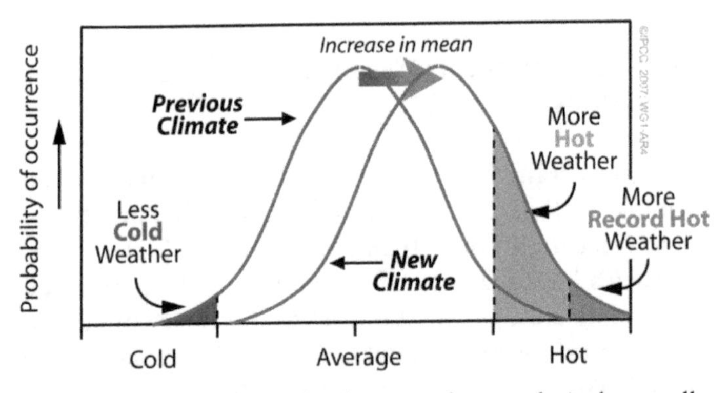

Figure 5.1. IPCC graph showing how relatively small increases in global average temperatures can lead to large increases in weather extremes.[3]

financial support and technical expertise. Norway, Sweden, and Denmark have thrown their support behind the effort on humanitarian grounds, going it alone within the European Union. (The German Green Party, part of the coalition government, has blocked any effort in Germany and, thus, at the EU level.) Everyone else, including the United States and China, is watching wearily, while India and Nigeria have voiced tacit support. So begins an all-out effort to design and launch high-flying planes within a decade.

The coalition is operating at a rapid pace, intending to launch the first flight within three years. While Airbus and Boeing decline to bid for rights to manufacture the plane, Bombardier, Embraer, and two smaller start-ups have offered their services. Embraer, supported by Brazil, wins the bid to custom-build a plane with enormous wingspan and large fuselage.[5] Brazil, now facing its own climatic emergency with large swaths of the Amazon burning, formally joins the coalition of nations. Several other companies, at first hesitant, join the effort. Rolls Royce supplies the aircraft engines.

A political emergency response rapidly gaining momentum

Scientists scramble to offer their insights, with some forming a "Red Team" focused on risks, while others, the self-declared "Blue Team," strive to guide the effort. Harvard's half dozen SCoPEx balloon flights, launched in the mid 2020s, still stand as the only set of stratospheric outdoor experiments. While calcium carbonate looked like the most promising compound due to its optical properties, the coalition decides to go a "safer" route and opts for sulfur dioxide, citing last century's Mt. Pinatubo eruption as the "natural" analog for its efforts.[6] (Filipino national pride appears to play a small but not insignificant role in the final political decision.)

While the coalition presses ahead, there are repeated meetings of the UN Security Council and, subsequently, the UN General Assembly. Both appear unwilling and unable to come to a plan of action – neither on immediate humanitarian measures nor on how to respond via mitigating climate change in the first place. (Some country delegations argue that solar geoengineering should be seen as a mitigation measure; most disagree.) By now the Philippines, Australia, and the three Nordic countries are joined by a handful of African and small island nations. Many others, led by Germany, formally condemn the move, while they privately express relief.

The coalition moves fast, commensurate with the emergency setting. Instead of dispersing 200,000 tons of SO_2 in the first year, increasing linearly by 200,000 tons each year, like in the slow ramp-up of the "rational" approach, the coalition plans on dispersing one million tons of SO_2 in the first year. That amount can be expected to reduce global average temperatures by around 0.1°C. The coalition is planning on doubling that amount in year two, increasing by one million tons of SO_2 for the first ten years. That pace makes many scientists nervous. But by now the decisions are largely political. Desperation begets radical action.

New Zealand has since joined the coalition. Its former prime minister, Jacinda Ardern, who distinguished herself during the Covid-19 crisis in the early 2020s, now leads The Elders, an independent group of global leaders helping guide the effort in an increasingly formal role. While things are moving fast – and not without opposition – the vow of radical transparency seems to pay off. It seems increasingly certain that global average temperatures will, in part, be managed by humanity. The coalition effort seems both radical and fraught, yet relatively stable.

6

Millions of geoengineers

The year is (still) 2035, or perhaps 2045, but nobody really knows when things started in the first place. It might have been a few dozen experimental balloons sent into the lower stratosphere. Or perhaps it was a covert rail guns project pursued by the Saudi military. The project baffled Mossad and CIA military analysts. The guns never seemed to point anywhere but straight up into the air. In retrospect, the mission seemed clear.

Early experiments focused on delivering tiny payloads – 5, 10, or perhaps 20 kilograms of SO_2 at a time – into the lower stratosphere. It began almost as a joke, with a few hobbyists buying balloons on Alibaba for $20 each. But instead of attaching a GoPro camera and hoping to retrieve the footage filming the curvature of the Earth when the balloon floated briefly in the lower stratosphere, the intent now was for the balloons to pop and release tiny amounts of SO_2.

The climatic impact: zero. The initial operation: undetectable. There was only one brief brush with fame, when a hobbyist filmed himself handling the small SO_2 canister as it exploded in his face, almost earning him a Darwin Award. The video went viral but was soon forgotten.

The Saudi experiments with marine rail guns had their own sets of mishaps. Most were kept secret or written off as military training accidents. It turns out that shooting small packages of a violent industrial gas 20 kilometers straight up presented a few challenges. Soon rockets entered the picture, in part guised as the Saudi's fledgling space program. The running joke among space watchers was that most Saudi rockets never actually made it into space. Many exploded well before. Few dared to guess that had been the intent all along, part of an experiment to learn how to deliver increasingly larger payloads of aerosols into the lower stratosphere.

High-flying planes are the most discussed solar geoengineering technique, and for good reason. They are widely considered to be technically feasible. They would also be relatively cheap to build, fly, and operate. Think single-digit billions of dollars per year at first, then perhaps ten times as much for an established deployment program.[1] And, even if it was $100 billion a year, global net benefits of a cooler planet might vastly outnumber those direct deployment costs. Few potential public policy interventions have as large a benefit-cost ratio – vaccination territory.

There are large potential risks that might prompt a responsible, rational actor to decide not to pursue solar geoengineering, but political actors aren't always all that responsible or rational. Concerns over a deployment program gone wrong – or simply *perceived* to have gone wrong – may outweigh concerns over not acting on global warming in the first place. The politics of blame avoidance are all too real.[2] But what if none of this mattered? What if the deployment decision was entirely out of the hands of elected officials?

Having profited handsomely from the carbon age, Saudi Arabia is in a lose–lose situation. It tops the list of countries with the most to lose under unmitigated climate change. Some parts of the country are already dangerously hot and predicted to get even hotter.[3] Meanwhile, it is also among the biggest losers in a world that commits to significant carbon emissions cuts. Yes, it, too, has dabbled in solar energy in the early twenty-first century, investing in a major solar project here and there. It even considered doubling down on a solar-only strategy – Norway-style: keep drilling oil and selling it abroad while trying to wean its domestic economy off fossil fuels altogether. That didn't last long.

Sensible climate policy contributed to the Saudis' desire to pursue that strategy. All three of the horsemen of the climate policy apocalypse were at work: the rebound effect, spatial leakage, and temporal leakage.[4] Fuel-economy standards passed in Europe and the United States made driving there more efficient. That, in turn, lowered oil demand wherever fuel-economy standards had been put in place, lowering global oil prices and, thus, increasing oil demand elsewhere. Temporal leakage – aka the "Green Paradox" – meant that the very prospect of stricter climate policy later on drove the Saudis and others to pump more earlier. Whom- or whatever was to blame for historically low oil prices, the overall effect was clear.

Saudi Arabia and a handful of allies in the region had since doubled down on pursuing a full-on fossil-only strategy. That meant pumping as much as possible as long as anyone was still willing to pay for oil and gas. All that put together meant that the Saudis and their allies had every incentive in the book to pursue solar geoengineering – and to attempt to do so covertly.

Covert at first and all-but unstoppable

Attempting to lower global average temperatures without anyone noticing is difficult, to say the least. Keeping it a secret forever is out. It's also not what the Saudis intend to do. Instead, they "simply" intend to create a decentralized system, one that is so metastasized that it is difficult, if not impossible, to kill entirely. Instead of lofting aerosols via specifically designed, high-flying planes, the Saudis, thus, have been looking to other technologies – from balloons to rail guns.

In doing so, the Saudis have seeded, without much notice, a large pro-geoengineering effort among "grassroots" environmental groups in the United States and a handful of European countries. Initiated through Saudi-funded social media efforts, the groups have taken on a dynamic of their own, counting hundreds of thousands of members. Their means: do-it-yourself, high-flying balloons designed to burst in the lower stratosphere and release 5 to 10 kilograms of SO_2 each.

Governments have been in hot pursuit, attempting to squash the efforts. The U.S. Federal Aviation Administration has since barred such balloon launches, but it, too, has difficulties stopping more than a small fraction of such flights. Most concerning, the Chinese Navy has recently discovered an autonomous ship off the coast of West Africa launching similarly small amounts of aerosols by using marine rail guns.

Overall, the Saudi efforts cost at least an order of magnitude more than a plane-based deployment program would, but cost effectiveness has never been the goal. And costs are still relatively low – relative, that is, to unmitigated climate change. The goal is to build a system that is all-but impossible to kill.

This deployment scenario occupies the opposite end of David Victor's "Greenfinger."[5] Call it "highly decentralized solar geoengineering" (Table 6.1). That's at least what Jesse Reynolds and I did in a political science paper thinking through just that possible deployment scenario.

A good analogy might be drug distribution, and global attempts to do something about it. The global annual supply of cocaine is roughly 1,500 tons, with some 70% of it originating in Colombia. The global annual supply of opium is roughly 10,000 tons, with some 80% of it originating in Afghanistan.[7] Either would fit onto a densely packed container ship or perhaps a freight train. But, of course, that is not how drugs are distributed, seeing as it is generally considered to be a crime to do so. There are entire UN agencies dedicated toward eradicating their distribution, with various drug-related treaties in place.

Table 6.1. Categorization of solar geoengineering deployment by type and number of entities involved in deployment.[6]

	Approximate number of actors deploying solar geoengineering			
Character of deployers	1	~10	~100	> ~1,000
State	Unilateral	Minilateral	Multilateral	n/a
Nonstate	"Greenfinger"	Moderately decentralized nonstate solar geoengineering		Highly decentralized solar geoengineering
Probable means of delivery	Newly designed aircraft (deployment costs ~\$1.4/kg SO_2)[a]			Small balloons (~\$5/kg SO_2)[b]

[a] Rough estimates suggest costs of around \$1,400 per ton of sulphur dioxide (SO_2) deployed, carried into the stratosphere in form of sulphur and burned *in situ*.

[b] At a cost of ~\$25–50 for a small balloon carrying ~5–10 kg of SO_2.

Despite near-universal agreement to eradicate illicit drug trade, only a very small number of rogue states is necessary to guarantee the continued supply of these drugs. Both the production and especially the distribution of cocaine and opioids are highly decentralized.[8]

Solar geoengineering deployment may well mimic this admittedly highly speculative scenario in some ways. Solar geoengineering is not as personally addictive as drugs are. But between highly motivated environmentalists and state actors, there may well be the drive to seek out such decentralized deployment options.

Scenarios caught between is and ought

The emphasis here – even more so than with the prior two scenarios – is on "highly speculative." Are any of these scenarios technically possible *and* economically feasible? Yes, they are, based on what we now know. Are they likely? No. Or rather: I don't know, and, I'd hasten to add, neither does anyone else. They surely are not likely in the sense of, say, a ⅔ chance of occurring, especially not as described here. I also wouldn't give either of the three scenarios a ⅓ chance each. But I do think that any real-world solar geoengineering deployment program may well approximate elements of each of these scenarios.

The biggest question, as is so often in policy advice and analysis, is the fundamental tension between collective "rationality" on the one hand and individual "rationality" on the other; what might also pass under various shades of "realism." In one corner is the "rational," idealized deployment scenario involving scientifically grounded policy advice based on transparent, sensible decision criteria. In the other is the real world, governed by political whims and forces that are better summarized by Richard Hofstadter's *Paranoid Style in American Politics*, or treatises like Niccolò Machiavelli's *The Prince* and vivid descriptions of the difficulties faced by political leaders of all stripes in enacting anything resembling rational policy.[9] Vested interests dominate. Path dependencies are crucial. Reform happens in fits and spurts. None of that is particularly rational, at least not in the sense most of us would typically use that term.

There is another, even more important tension between what *is* and what *ought* to be. There may well be disagreement on what *is*, and there typically is such disagreement. But the real disagreement, of course, comes with the normative *ought*. Even if everyone were to agree on what is and will – would – be, it is nigh impossible to agree on what the world *should* look like.

This seems like a rather obvious point, but it must still be said. David Hume's missive that it is never possible to derive an *ought* from an *is* applies here, too.[10] The *ought* comes with preconceived notions about climate change, climate policy, and any possible role for solar geoengineering. It similarly comes with important value judgments.

Some of these value judgments are indeed better described as judgment calls over the evidence itself. How should a particular study be interpreted? Do certain risks matter? How should unknowns and unknowables be dealt with in making a (rational) decision?

Then there are some judgment calls that cannot be solved by looking up an answer somewhere, where there is no single underlying truth. Green moral hazards may top that list of value judgments vis-à-vis (solar) geoengineering. How one thinks about them may well determine whether one might consider solar geoengineering as the most likely on the one hand or as the most desirable on the other. The next chapter will be all about just this theme.

Part III

Governance

7

Green moral hazards

The first thing to know about moral hazard is that, more often than not, the term is woefully misused.[1] The dictionary definition is clear:

> mor·al haz·ard [ˈmôrəl ˈhazərd, *noun*]
> Lack of incentive to guard against risk where one is protected from its consequences, e.g. by insurance.

There are indeed plenty of examples where the term does apply directly. Think health insurance, where the knowledge that a doctor will attempt to fix things later might lead to riskier behavior now.[2] Or think about a direct technological intervention, like condoms or seatbelts. Using either might lead to some riskier behavior. Part of that reaction might be perfectly rational. Driving faster, after all, has its advantages. If seatbelts make driving faster safer, then doing so might indeed be perfectly rational for the individual driver, and the introduction of seatbelts as a matter of policy might be a perfectly sensible step for society.

All too often, though, the concept of moral hazard is misapplied, in two ways. One is that not every adverse reaction can be ascribed to moral hazard. There are, for example, plenty of other horsemen of the climate policy apocalypse. The energy efficiency rebound effect comes to mind, which implies that energy efficiency standards making cars, say, more fuel-efficient will lead to more driving and, thus, more energy use.[3] Others are spatial and temporal emissions leakages, whereby policies introduced in one particular place and time might lead to emissions increases elsewhere. Some of those emissions increases may indeed, at least indirectly, be linked to moral hazard-style shifts in behavior. Much of it is not.

Moral hazard in the context of geoengineering is no different. There, the concern is that the potential use of either carbon removal or, perhaps more so, solar geoengineering might lead to increased emissions.[4] The concern as such is both broader than moral hazard – any form of "crowding out" one action by another – and more specific. Instead of moral hazard, it is really "mitigation deterrence" or outright "mitigation obstruction."[5]

That possibility might extend to the mere mention of either technology: Think that a relatively cheap technofix might be available one day? Ease off the necessary emissions reductions now. That concern is well founded. Newt Gingrich penned the op-ed saying as much in 2008, during the ultimately futile fight in U.S. Congress to pass comprehensive climate legislation.[6] Then again, that's exactly what Gingrich would say.

Moral hazards = politics

The second thing to know about moral hazards is that, more often than not, invoking them is intensely political. But politics, here, comes with a twist. Typically, the political Right invokes moral hazards by pointing out how policy might shield individuals from their moral failings and encourage amoral behavior. That goes for health insurance in general as well as for more specific policies and technologies. Moral hazards, of course, are typically only applied to disliked policies. Universal health care? Moral hazard! Federal bailouts of favored political constituents? Silence.

Birth control interventions, including abortions, might be the most controversial – and, thus, political – example. Their mere availability, the argument goes, makes unprotected sex less risky and, thus, encourages more of it. (Full disclosure, it is also an example that hits home rather directly, with my wife working as a gynecologist at N.Y.U. Langone Health and as the division director of reproductive choice at Bellevue Hospital in New York.) One constant moral and political question then is whether to advertise and promote or whether to suppress any particular technology. Many on the Right would rather suppress abortions and certain types of birth control, declaring their very use immoral.

With geoengineering, it is the environmental Left who are worried about moral hazards. And this does not only apply to geoengineering. It extends to most anything where a mere technofix does not go far enough and might preempt more fundamental, complex societal changes:

> green mor·al haz·ard [grēn ˈmôrəl ˈhazərd, *modified noun*]
> Lack of incentive to fix deeply seated, complex environmental problems because of the possibility of a quick technofix, e.g. geoengineering.

Such worries may not stop with technologies alone, nor with geoengineering. Even comprehensive carbon pricing might fit the bill as an intervention that "merely" leads to lower CO_2 emissions, while not addressing all sorts of other environmental and societal problems. Yes, a sufficiently high price and other climate policies might "stick it to CO_2," but they do not "stick it to the Man" at the same time, at least not as much as some on the political Left might hope.[7]

There are indeed real arguments to be had about the best environmental policies and technologies. By now, for example, there is broad understanding that carbon pricing alone would not adequately address climate change.[8] Comprehensive climate policy ought also to include investments in low-carbon technologies in the form of direct subsidies.

There are similarly real arguments to be had about how to pass the most ambitious set of climate policies. One strain of this argument is whether to focus on climate policy in isolation, or to see it as an issue intimately linked to other, broader societal concerns. Policy sequencing matters, too. Hardly any country or jurisdiction has passed carbon pricing in isolation. Doing so typically goes hand-in-hand with – or follows – policies focused on subsidizing low-carbon energy or infrastructure more broadly.[9] All these are important active policy debates.

Now enter geoengineering into the mix. It's easy to see how some environmentalists would be concerned that any focus on carbon removal or especially solar geoengineering might detract from the need to cut CO_2 emissions in the first place: *Finally, we have some momentum on [much-needed policy X]! Why detract from it by talking about geoengineering?*

Moral hazards throughout environmental history

Geoengineering is hardly alone in invoking feelings of strong moral hazard. It is not even unique within recent climate policy discussions. Carbon removal is now roughly where talk of adaptation was in the mid 1990s, solar geoengineering not yet even that far. Back then, mere talk of the need to adapt to climate change already in store was often dismissed as a distraction from carbon mitigation in the first place. Even Vice President Al Gore was on record saying as much.[10]

Adaptation by now has a well-established place in comprehensive climate policy. Now, though, to a lesser extent carbon removal and more so solar geoengineering are often the odd ones out. The history of moral hazard in environmental thought, meanwhile, goes back much further.[11] By some accounts, modern environmentalism's very existence is intertwined with green moral hazard logic. That logic, in turn, comes in two forms. One is a principled rejection of any and all "technofixes" to environmental problems.[12] The second, intertwined with the first, is a rejection of any policy that falls short of a broader social revolution with widely held notions of justice at its core.

To be clear, I subscribe to many of these ideals. Rampant inequality, systemic racism, and all sorts of other societal ills clearly demand much broader, fundamental reforms. But there is a rather big step between arguing how a simple technofix can't solve it all – it usually can't – and how said technofix should never be touched because it doesn't solve it all. Most interventions – technological or otherwise – involve tradeoffs.

The two stories most often told in this context are of whale oil and horse manure.[13] Both stories played out in the mid to late nineteenth century and the early twentieth. Both were, at first, seen as clear-cut success stories of technological innovation and entrepreneurial spirit. Both also involve fossil fuels saving the day. Not least because of that, both stories have since become highly politicized.

Looking at the whale-oil story, for example, Matt Ridley, the self-described *Rational Optimist*, whose climate pronouncements often border on willful ignorance or worse, likes to reference Warren Meyer, a self-declared climate skeptic, by saying, "Greenpeace should have a picture of John D. Rockefeller on the wall of every office."[14] That quote serves several purposes, not least it appears to turn environmentalists' morality against themselves. See, if it hadn't been for the discovery of rock oil, the argument goes, the world would have continued to hunt right whales for their blubber aka whale oil. The truth, it turns out, is quite a bit messier, and it took Italian chemist Ugo Bardi looking through a wealth of whale-oil extraction and pricing data to piece it together.

North Atlantic whale-oil production peaked before 1850, well before rock oil became available in 1859. By the time kerosene became commercially available, whale-oil extraction was less than half its pre-1850 peak, leading Bardi to conclude, "the peaking and the initial phase of decline of whale-oil production were not caused by the availability of a better technology, but by the physical depletion of the resource."[15] The U.S. Civil War, too, played a role here, as it partially grounded the whaling fleet during the early 1860s. Moreover, kerosene lamps, revolutionary as they were in their own right, would soon be replaced by another technological leap. Thomas Edison invented the incandescent lightbulb by the late 1870s, a technology that would rapidly replace oil lamps and democratize access to artificial light on both sides of the Atlantic over the following decades.

So, first, it wasn't rock oil that led to the initial 50% drop in whale-oil supply. That was due to woeful overexploitation of right whales. And second, instead of crediting Rockefeller and fossil fuels with serving as a backstop technology of sorts, it might be more accurate to credit Edison and the electric light for doing so. All that doesn't just put a jinx on stories of fossil fuels as an environmental savior, it also complicates how this story fits within the green moral hazard framework. New technologies often come with pluses and minuses. Sometimes the overall picture is indeed mixed or dominated by negatives. Sometimes new technologies might do a lot of good. The result often depends on broader societal – political – choices.

The horse-manure story paints a similar – and similarly complex – picture. The setup is one of fossil-fueled technological salvation:

> While whales were declining, horses multiplied – not out in the wide open plains but right in the heart of major cities. Horses had been the most important mode of transportation for thousands of years. It wasn't until the late 1800s that they really took off. New Yorkers made over 30 million trips on horse-drawn carriages in 1860. By the end of the decade, that figure topped 100 million, and there was no end in sight. In the 1890s, 200,000 horses produced 2,500 tons of manure a day. It took thousands of horses just to haul the manure away. Much of it was simply dumped into empty lots. One doomsday scribe infamously predicted that by 1930 horse manure would blanket Manhattan three stories deep. In 1898, the world's first urban planning conference broke up in disarray after only three of ten scheduled days. Delegates could not see a way out of the gridlock and stink. Mobility was as essential and addictive back then as it is today. It was a transportation, environmental, and public health nightmare rolled into one. The model of cities looked wholly unsustainable.
>
> We know what came next. John D. Rockefeller, Henry Ford and others saved the day with oil-powered cars that pushed horse-drawn carriages out of New

> York. In 1900, cars were still a luxury turning heads. That year 4,192 cars sold in the entire United States. By 1912, sales topped 350,000, and cars for the first time outnumbered horse-drawn carriages in New York City. Five years later, the last horse-drawn streetcar was retired for good and by the 1920s, horses were all but gone from city streets save for horse carriages shuttling tourists through Central Park and the occasional mounted cops, who to this day roam through Times Square and Central Park.[16]

Yet, once again, the story isn't quite as simple. Yes, cars provided a superior technology compared to horse-drawn carriages. They were faster, safer, and cleaner – at least to the extent that the tons of CO_2 and other pollutants coming out of their tail pipes tended to dissipate instead of accumulating like horse manure. But it wasn't nearly as simple as most versions of the story seem to suggest: a triumph of technological innovation and entrepreneurs – capitalism – over environmental despair. Far from it, though the lesson here is different from that of the whale-oil story. It is about the importance of government policy and public investment – from interstate highways to Navy ships patrolling the Persian Gulf: "When General Motors bought up streetcars and converted them into buses through a wholly owned subsidiary, it effectively destroyed mass transit in over forty cities throughout the United States between the 1930s and 50s."[17] New technologies don't appear in a societal or political vacuum. Some might at first appear to grow into a void, but even the internet famously depended on heavy public investment.

These decidedly mixed histories make many environmentalists' skeptical reactions to geoengineering technologies more than understandable. An even clearer example is nuclear technology, pitting Promethean promises directly against nature. The advent of the nuclear age also coincided with the birth of modern environmentalism.[18] Rachel Carson, Barry Commoner, and others central to the founding of the modern environmental movement pointed to nuclear fallout as the clearest example yet of how, in Commoner's words, "we tend to use modern largescale technology before we fully understand its consequences."[19] Carson pointed to nuclear fallout to explain the dangers of DDT.[20]

It was against this backdrop that (solar) geoengineering was introduced, and it was at first quite deliberately presented as a "technofix." The 1965 report to President Johnson, for example, presented solar geoengineering as the sole response to climate change, very much reflecting this technofix mentality.[21] This presentation as a technofix all but guaranteed that geoengineering would be broadly understood as a moral hazard.[22]

There was pushback from the very beginning. Even before the first Earth Day in 1970, Stewart Brand declared in his *Whole Earth Catalog*'s statement of purpose, "We *are* as gods and might as well get good at it."[23] Brand offered a techno-optimistic vision of environmentalism, one that is pro-nuclear power, pro-genetically modified organisms, and pro-geoengineering – not blindly so, but clearly as part of the overall solutions package. In fact, Brand participates in these debates to this day.[24]

I can see the appeal of this techno-optimistic brand of environmentalism. (Full disclosure: I once shared a stage with Brand and David Keith at a conference organized by the MIT Media Lab discussing solar geoengineering research.) But despite Brand's prominence, it was clearly a minority view then and it remains one to this day. It has often also been coopted by self-declared "rational" or "pragmatic" environmentalists who seem to take pride in controversy and in being seen as either having left the environmental movement or in never truly having been a part of it.

(Solar) geoengineering and moral hazard

Both carbon removal and solar geoengineering occupy a rather complex role in this debate. Neither technology makes sense even to contemplate without a clear understanding that climate change is sufficiently bad to potentially warrant the intervention. That still raises plenty of scientific and ethical questions. One way to encapsulate that debate is to ask the deceptively simple question of whether (solar) geoengineering should be seen as a first or last resort. The correct answer may well be neither. In fact, I would emphatically argue that it is. But looking at either extreme position is still instructive.

To those few calling either carbon removal or especially solar geoengineering a first resort, it might seem unimaginable to wean the world off fossil fuels. The 1965 U.S. presidential report (implicitly) took that position. Newt Gingrich's 2008 op-ed did the same. Why pass climate policies aimed at decarbonizing the economy if there is a simple technofix? That thinking, once again, directly invokes solar geoengineering as a green moral hazard. It seemingly avoids any of the harder, more expensive steps necessary to address climate change. I would describe any such position as wishful thinking, willful blindness, or worse.

Treating either carbon removal or especially solar geoengineering as a last resort, meanwhile, might be just as dangerous. What if it doesn't work as advertised? And even if it does, solar geoengineering is no solution. Carbon removal at least addresses the proximate root cause. Solar geoengineering doesn't even do that. (And I'm saying "proximate" here because carbon removal does not, in fact, address the wider root causes those concerned about moral hazard might point to.) Solar geoengineering would indeed show faster results than cutting CO_2 emissions or removing it once in the atmosphere; but fast is not the only objective.

The only sensible way to approach either carbon removal or solar geoengineering, then, is not as an either–or but as a yes–and. Neither should stand on its own but must instead be seen as part of a much broader climate policy portfolio that includes cutting CO_2 emissions in the first place, as well as plenty of adaptation to what's already in store. That also makes any geoengineering deployment scenario neither a first nor a last resort.

Of course, saying so will not make it so. While there are indeed some small rational tradeoffs for all four interventions – mitigation, adaptation, carbon removal, and solar geoengineering – moral hazard is not about what would happen in an idealized, rational scenario. Moral hazard worries around (solar) geoengineering are all about what would happen in the irrational, fickle world we inhabit. That holds true for carbon removal but even more so for solar geoengineering.

Given the "free-driver" properties of solar geoengineering leading to the "not *if*, but *when*" conclusion around its eventual deployment, geoengineering governance ought to focus on channeling these fundamental forces into a productive direction. To a first approximation, this implies simply suppressing them – globally. (See the next chapter, calling for a deployment moratorium now.) Given how the moral hazard of (solar) geoengineering cannot be wished away, it also means channeling the moral hazard forces themselves in the right direction.

Moral hazard and its inverse?

It's clear that (solar) geoengineering as a technofix implies a tradeoff with the desire to cut CO_2 emissions in the first place. Or does it? The empirical evidence is decidedly mixed. When we scoured the academic literature about five years ago, we found some thirty surveys attempting to elicit public opinion on solar geoengineering's moral hazard.[25] Most of these surveys concluded that there is some kind of moral hazard. The mere availability of solar geoengineering, it seems, would encourage survey respondents to want to ease off CO_2 emissions cuts. Alas, these surveys also often raise more questions than they answer.

One concern is that generally very few survey respondents know what solar geoengineering is before taking the survey. That's not an insurmountable challenge *per se*. Surveys often try to elicit basic familiarity with some obscure (or not so obscure) topic, scientific or otherwise. "Have you heard about the SARS-CoV-2 virus?" "Which branch of the U.S. government makes laws?" The fraction of respondents showing some basic familiarity with the subject might itself be revealing, one way or another. The likes of Jay Leno and Jimmy Kimmel have been milking admittedly biased versions of such tests for laughs for years.

The public's unfamiliarity with solar geoengineering is revealing in itself. But surveys typically don't stop with this finding. They then go on to elicit a number of other responses, including around moral hazard. A lot then depends on how the information is presented. Describing it as a risk-less, cost-less, innocuous intervention would elicit very different reactions than a description invoking images of madman scientists taking over the world's climates on behalf of fossil-fuel interests. Real framing choices, of course, are much subtler. One that's rather obvious in hindsight but was apparently ignored by all such prior surveys is what's typically called acquiescence bias.[26]

Especially when survey respondents don't have strong feelings either way, it turns out they have a tendency to agree with the way the question is phrased. Ask whether people think solar geoengineering might detract from the need to cut CO_2 emissions, and to someone who had not previously thought about the question, it appears like a reasonable possibility. Ask the same person whether they think the availability of solar geoengineering might *encourage* more CO_2 emissions cuts, and they consider that a reasonable possibility, too. The way the moral hazard question is asked matters. That finding alone doesn't invalidate any and all prior surveys pointing to moral hazard, but it should certainly add a hefty dose of skepticism.

Another dose of skepticism is in order, since these surveys indeed just elicit answers to hypothetical questions. There are better and worse ways of doing that – see "acquiescence bias" – but there's no escaping the fact that, in the end, it's "just" a survey. Ask a silly – or biased – question, get a silly – or biased – answer. Potentially one better: observe how people actually behave.

Social scientist Christine Merk did just that with two of her colleagues in the first-ever revealed-preference survey of around 600 test subjects. Instead of asking her test subjects to answer hypothetical questions around moral hazard, she let them vote with their feet, or rather, their Euros. The experiment, in rough terms: Split 600 participants into three groups. The first two groups are told about climate change in varying detail. The third is told about climate change and the possibility of solar geoengineering in the form of stratospheric aerosols. All three groups are paid for participating in the experiment, with one twist: They can choose how much of their survey payments they'd like to use to pay for carbon offsets. (All that raises further questions about the effectiveness of carbon offsets as a response to climate change, but let's leave that for another day.[27] It should not bias the results here one way or another.) The resulting paper's title summarizes the findings rather modestly: "Knowledge about aerosol injection does not reduce individual mitigation efforts."[28]

I'd go one step further: Merk and her colleagues' experiment shows the very real possibility that telling the uninitiated about solar geoengineering might indeed *increase* their desire to cut CO_2 emissions. The direct comparison of the effect of telling 200 subjects about solar geoengineering versus just telling them more about climate change itself shows just that. It's solar geoengineering that has the additional motivating effect, not reading more about climate change in general. Why would that be?

One hypothesis for this *inverse* moral hazard finding is that solar geoengineering sounds so scary that it makes climate change itself look like more serious of a problem: "Wait, serious scientists are talking about doing *what*? Perhaps there's something to this whole climate change thing after all!" To be clear, that's just one hypothesis, and Merk's study does not show conclusive evidence for or against this hypothesis. (Full disclosure: Christine has since become a co-author of two essays on moral hazard, and we have been running extensive online experiments to try to get to the bottom of this very question.)[29]

Moral hazard is a clear possibility. So is its inverse. It all seems to depend on how things are presented, who is asked, and, as usual, on the broader context. That raises the all-important question of whether any of this actually matters. The public might think one way or another. But what about those who end up making the decisions? Structural biases and deeply seated, vested fossil-fuel interests surely matter more. It is indeed just that type of "moral hazard" that should worry us. Fossil-fuel interests advocating for either carbon removal or solar geoengineering solely as a way to delay cutting CO_2 emissions would likely be the most consequential type of intervention.

That kind of fossil-fuel advocacy might indeed take several forms. One might be classic lobbying of politicians. Another might well involve attempts at misleading the public by introducing either carbon removal or solar geoengineering as the easy way out of our climate policy dilemma: "Don't listen to those environmentalists telling you that you can't drive your gas guzzler." The real fossil industry actions, of course, would be much subtler and, thus, dangerous. For example, given the general public's low familiarity with carbon removal and especially with solar geoengineering, part of vested interests' advocacy may well be focused on trying to shape a positive view of either set of technologies, downplaying any risks.

Education, education, education

There's no silver bullet to addressing any of these concerns. At the very least, it is important for everyone involved in geoengineering and climate conversations more broadly to be aware of them. That also means having more inclusive conversations in the first place. Education is key. The goal is nothing short of elevating the climate conversation by (also) including carbon removal and solar geoengineering in their proper place.

I realize I'm preaching to the choir here. Having made it this far into a book on geoengineering alone speaks volumes. The overall goal, of course, must be much broader. One part of this equation, for example, is having balanced takes on geoengineering in climate books that aren't focused on geoengineering itself but paint a much broader picture of climate change and climate policy. I, for example, joint with the late, great Marty Weitzman, have tried to do that in *Climate Shock*, a book that is primarily about low-probability, high-consequence climate impacts.[30] I am, of course, far from alone.

The most prominent recent example might be Eric Holthaus's *The Future Earth*.[31] The subtitle alone says a lot about Holthaus's general outlook: *A Radical Vision for What's Possible in the Age of Warming*. Holthaus goes decade by decade through a "radical" yet hopeful vision of the world. That vision, for example, includes an entirely decarbonized U.S. economy by mid century. Despite all that, and not unlike the "'rational' climate policy" scenario in Chapter 4, Holthaus's vision includes a preview of solar geoengineering – and certainly not because he likes the idea. Nobody should. Despite conceiving of a world that includes plenty of seemingly radical decarbonization technologies and policies, he envisions solar geoengineering playing a certain, limited role.

For every *Future Earth*, of course, there are many other climate books that portray geoengineering in anything but a balanced fashion. Some are naively optimistic about carbon removal and solar geoengineering. Many more appear to be openly hostile. And broadening the conversation, of course, doesn't end with books or erudite writings.

Hollywood, as usual, plays an outsized role. Sadly, there, "education" is hard to find. (Solar) geoengineering is simply all-too tempting to portray through some kind of gee-whiz lens. The movie "Geostorm" might be the most prominent such example, with a globe-spanning satellite network controlling the world's weather. (Narrator: No, that is not a sensible solar geoengineering technology.) I'm also ignoring "Snowpiercer" here, which has a failed solar geoengineering attempt that instead created Snowball Earth as backdrop, but really plays out aboard a train circling the globe, cannibalism and all.[32] (Narrator: The movie makes for some morbid entertainment, but it, too, is not a good introduction to solar geoengineering.)

Meanwhile, the first sci-fi novels have come out, portraying solar geoengineering in relatively realistic terms, while helping their readers think through some of the more profound social and political implications. Eliot Peper's *Veil* might be the best example in this genre, Kim Stanley Robinson's *Ministry of the Future* the most prominent.[33] Peper's story is replete with a solar geoengineering deployment scenario largely grounded in reality, a sinister oil conglomerate SaudExxon selling the Earth for profit, and still plenty of humanity to contemplate how solar geoengineering might play out on a planet struggling to bring global warming under control.

None of this guarantees an elevated (solar) geoengineering conversation. It also does not remove the need for deliberate, public engagement processes like that of the external advisory committee for Harvard's SCoPEx project, nor does it remove the need for other attempts to educate and broaden the conversation with policymakers and the public.[34] That is especially true in light of much more deeply seated problems. Technological choices and technology policies are not typically egalitarian, democratic, or pluralist, especially when viewed through a global lens.[35]

Yet it is precisely an imaginative portrayal like in Peper's *Veil* that might lead toward the kind of rapid, deliberate education that is necessary to have sensible discussions around (solar) geoengineering and its role within a wider and widening response to climate change – green moral hazards and all.

8

Research governance

Legal opinions differ on the applicability of current laws and treaties to solar geoengineering. Given the dearth of precedent, a lot is up to interpretation and outright speculation. The informed legal consensus suggests one thing: There is no law or treaty that would ban or even comprehensively regulate solar geoengineering (or carbon removal, for that matter).[1] That moves solar geoengineering away from any definitive legal pronouncements and instead moves it toward broader "governance" conversations.

"Governance" here is essentially short for: "We need to talk." That's not a bad starting point, given where things stand. It's also a recipe for just that: lots of talk. There are research conferences, panels, dialogues, conversations, exchanges, and assorted other meetings large and small. Some of the talk is happening among climate and other natural scientists; lots of it is happening among social scientists and civil society organizations. More and more talk is spanning various realms, disciplines, and national borders.

I'm not saying any of this to be facetious. Quite the opposite. It is simply the sign of a burgeoning area of scientific inquiry with enormous potential impact and societal implications. And all of this talk is beginning to bear fruit. One set of important conversations is happening among what the British would appropriately call the learned societies – including those in the U.K., the U.S., China, and the World Academy of Science, focused on developing countries.[2]

The U.K. and German governments, meanwhile, have been funding research initiatives. China's has begun to do the same, and, in December 2019, the U.S. government, too, set aside its first $4 million for a research program led by NOAA, the National Oceanic and Atmospheric Administration.[3] It's early days, but seeing as budget *is* policy, even $4 million for research is an important first step.[4] The National Academies' call for $100 to $200 million in funding over five years here is a real sea change.[5]

There is also lots of conversation about whether the topic should be researched at all. My bias there is clear. It should. If nothing else, the "not *if*, but *when*" logic leaves us no other choice. (If you have any doubt, see "An unapologetic call for research" on page 139.)

A moratorium now

But there is a big step between *research* and actual *deployment*. The distinction, of course, only becomes relevant for research conducted outside of a computer model or lab. With any potential outdoor experiments, the line is at once complex and fluid, and it is also rather easy to draw in principle.

The two crucial criteria are intent and effect (Table 8.1). If the intent is to conduct an outdoor experiment

Table 8.1. The line between solar geoengineering research and deployment depends on intent and effect.

Intent \ Effect	*De minimis* effects[a]	Measurable temperature (or other) impacts[b]
Intent to research	Research	(Misguided research)
Intent to alter temperature (or other variables)	(Failed deployment)	Deployment

[a] For example, for changes in annual average radiative forcing of less than 0.000001 W/m^2.[6]

[b] For example, for changes in annual average radiative forcing of more than 0.01 W/m^2[7]

and the effect is so small that it does not have any direct measurable impact on the surrounding area, it's research. If the intent is to have an impact, and the action is indeed large enough to have a measurable temperature (or other) global or large regional impacts, it's deployment.

The other two squares are a bit murkier, but neither seems like a smart intentional choice. Action with the intent to alter temperatures but too small to have an impact would simply be failed deployment. Action with the intent to research but with measurable temperature impacts might be best described as misguided or unnecessarily ambitious research, at least for the time being – and that is precisely my point.

One potentially productive way forward in solar geoengineering research is to institute an effective moratorium on any and all deployment of solar geoengineering, ideally a legally enforceable ban for any intentional deployment of research above a certain size. Here, too, are questions aplenty: Who implements it?

Who might comply with it? Who might not? Who could modify it, or lift it altogether, and what are the criteria for any of these steps to happen?

I'm far from the first to call for such a moratorium. That distinction goes to none other than the late Ralph Cicerone. A renowned atmospheric scientist, he was the president of the U.S. National Academy of Sciences when he wrote an essay in support of solar geoengineering research to accompany Paul Crutzen's famous 2006 thought experiment about trading off stratospheric sulfate aerosols with tropospheric ones.[8] Cicerone proposed a moratorium on any deployment activities, all while enabling and even encouraging research on the topic.

The proposal received some attention at the time, including in the runup to the March 2010 Asilomar conference.[9] However, a good indication of the, at times, glacial pace of research governance conversations is the title of a 2013 essay by Ted Parson and David Keith: "End the deadlock on governance of geoengineering research."[10] Several years have passed since then – and with it many more conferences, panels, dialogues, conversations, exchanges, and other meetings. Talk of an outright moratorium has, in fact, found some takers, though, I might add, rather disingenuous ones.[11]

The small yet vocal anti-technology ETC Group has repeatedly pointed to the Parties to the Convention on Biological Diversity as having issued a "de facto moratorium" on (solar) geoengineering research. It has not. In 2010, the Convention issued what is best described as a "normative framework for considering the use of geoengineering."[12] That's a mouthful. It might also be an appropriate way for an international institution to approach the topic.

The high-level language leaves plenty of room for interpretation, though instead of a "de facto moratorium," the Convention's Parties issued a direct call for more research. In 2016, they issued a call for "more transdisciplinary research and sharing of knowledge among appropriate institutions ... in order to better understand the impacts of climate-related geoengineering on biodiversity and ecosystem functions and services, socio-economic, cultural and ethical issues and regulatory options."[13]

It is clear that this research ought to proceed only if it has a negligible – *de minimis* – effect on climate and biodiversity alike. It is similarly clear that it is no *carte blanche* for all research, but it *is* a call for more research. All that is indeed what one might hope to hear from such an international body. It might be even better for this type of language to emerge from a broader global body, like the UN Environment Assembly or the UN General Assembly, but the Parties to the Convention on Biological Diversity appear like a good second-best.

An unapologetic call for research

The flipside of a deployment moratorium now is indeed an explicit permission for research efforts to proceed, much like the 2016 call by the Parties to the Convention on Biological Diversity. While the time for deployment is much later – or maybe, in a rational world, never – the time for research is now. But why research? A quest for knowledge for the sake of creating knowledge is not sufficient here. Not all knowledge is good. Creating smallpox – or another coronavirus – in a lab, especially if done with the intention to harm indiscriminately, is clearly bad.

The case for solar geoengineering research, then, ought not to rest on some principled philosophical stance about knowledge creation. It is much more practical than that, firmly grounded in reality. First, there is the harsh climate reality. The last time concentrations of carbon dioxide were as high as they are today was over three million years ago, during the Pliocene. Global average temperatures were at least 1°C higher than they are today. Global average sea levels were at least 10 to 30 meters higher.[14]

All that calls for much more urgent climate action. That may indeed be the largest role for solar geoengineering research: not for its own sake, but as a wake-up call for broader climate action. It also counters arguments around the "moral hazard" of solar geoengineering – how it might detract from the need for more urgent climate action.[15] Quite the opposite: calling attention to solar geoengineering ought to increase the overall urgency for climate action more broadly.[16]

Second, there is the argument that solar geoengineering itself may indeed do some good. If we already knew, more or less with certainty, that solar geoengineering would be unequivocally bad, that argument would be moot. I would strongly argue that we simply do not know enough at this point.

Saying that solar geoengineering might do some good is sometimes misconstrued as some kind of "gotcha" moment: "Hah, I knew it, you aren't just for more research. You are *for* solar geoengineering." Not so. Arguing for solar geoengineering research on the basis that the technology may indeed do some good is distinct from saying that solar geoengineering itself *is* a good idea. It may be, or it may not. We simply do not know enough – hence the need for more research.

The stronger argument *against* research is one that says researching a new technology is a self-fulfilling prophecy. The more research infrastructure there is, and the more institutions are built to guide such research, the more likely it is for the research to continue toward actually deploying the technology. That argument clearly holds some credence. After all, institutions, once established, are hard to kill. Saying that this case is somehow different would not be convincing.

In fact, it is precisely this case for institutional lock-in that argues for research programs based at universities, and not, for example, anywhere within the established government bureaucracy. (Alan Robock was rightly concerned about the CIA contacting him about geoengineering.)[17] Any kind of research must be open, transparent, and generally go above and beyond best-practice principles of responsible research and innovation.[18]

In the end, though, the case for solar geoengineering research comes back to the fundamental property of stratospheric aerosols being fast, cheap, and (highly) imperfect. The last part calls for research to find out more about the risks and uncertainties. The first two imply that – research or not – we may be slithering toward deploying solar geoengineering hastily, in some kind of (perceived) climatic emergency.

There is little that researchers themselves can do to alter these fundamental properties, other than, of course, unearthing that they may simply be false. That, too, would be incredibly valuable to know, and sooner rather than later. Without any such findings, though, we are back to the geoengineering gamble, staring down "not *if*, but *when*" – research or not. Research or not, another bit is clear: While the decision on research itself should largely be in the hands of researchers – with learned societies, international bodies, and others helping guide the questions as well as the open and transparent dissemination of knowledge gained – any deployment decision must be out of the hands of researchers.

Any deployment decisions can – must – only be taken by well-informed elected leaders. A deliberate moratorium now may well be the best way to create the kinds of conditions that will enable such a decision process to be put in place.

Epilogue: The inevitable gamble

"We have 12 years to cut emissions to zero."[1]

Versions of this statement are all around. They are at once wrong in their specificity and oh-so right in the overall sentiment they convey. It is well past five minutes before midnight in the climate emergency, to use another cliché. The "shock" in *Climate Shock* is real.[2]

If only the world had acted after that pivotal Jim Hansen testimony before the U.S. Senate Committee on Energy and Natural Resources on June 23, 1988, which led the *New York Times* to announce on its front page that, "Global warming has begun, expert tells senate."[3] If only it had acted when Exxon scientists confirmed the role of fossil fuels in the late 1970s,[4] or when President Lyndon B. Johnson's Science Advisory Committee delivered its report in 1965.[5] Exxon knew. Lyndon B. Johnson knew. Svante Arrhenius knew – in 1896.[6] *Popular Mechanics* ran an article in 1912 discussing the greenhouse effect and the role of burning coal in raising temperatures![7]

If only.

Instead, we are staring down a world where the true price of each ton of CO_2 emitted is a multiple of what the vast majority of the world is apparently willing to pay.[8] The risks of unmitigated climate change are all-too visible. The drive to a quick technofix solution is all-too real.

What often keeps me up at night – quite literally, frankly – is the fear that we might be slithering toward deploying solar geoengineering without having done the hard work. That we – us researchers – are missing something fundamental – and that time just isn't on our side. It took a quarter century after the eruption of Mt. Pinatubo lowering global average temperatures by 0.5°C for that first *Nature* cover article to estimate the agricultural effects of scattering sunlight.[9] A quarter century.

Meanwhile, people are dying from unmitigated climate change today! The clear answer, of course, is to mitigate: cut CO_2 emissions, now. That may well be the best use of solar geoengineering today: scare people into wanting to mitigate more. (See moral hazard, *inverse*).[10] But what if deploying solar geoengineering, arguably another form of mitigation, might indeed save more people sooner?

Put slightly more philosophically: At what point did not cutting enough CO_2 turn from an error of omission into an error of commission? If we believe we've crossed that threshold – and I certainly do – at what point then does something similar apply to geoengineering?

That is precisely where we return to the "gamble" inherent in solar geoengineering. Pursuing it is risky, perhaps unduly so. *Not* acting is similarly risky, perhaps even more so. No simple benefit-cost analysis will tell us which way to go. The decision is all about risk–risk tradeoffs, putting the risks of unmitigated climate change against the risks of potentially pursuing solar geoengineering.

That's a highly uncomfortable position to be in. It's a gamble, and a planetary-scale one at that. It's also a gamble we aren't being *asked* to participate in, or perhaps to observe as neutral spectators. It's a gamble we're being pushed to play: "not *if*, but *when*." I for one would much rather have us be prepared when that time comes.

And yes, we may learn along the way that deliberately deploying solar geoengineering could be a good idea, making us – and a big question is who that "us" is – more eager to engage in the gamble in the first place.

None of this is comforting. None of it should make us – *anyone* – feel good about wanting to deploy solar geoengineering, or even to research it. We should never become numb to this feeling about engaging in such a massive potential gamble. The one thing we can do is try to be prepared, much better than we currently are.

That includes all of us – researchers, policymakers, environmentalists, anyone involved in climate policy, and the engaged public. There are still many more questions than definitive answer. Perhaps there always will be. But we do need to do the work to be prepared.

I often think about President Barack Obama describing to Jerry Seinfeld in the basement of the White House during an episode of "Comedians in Cars Getting Coffee" how politics is most like American football: "A lot of players. A lot of specialization. A lot of hitting. A lot of attrition. But then, every once in a while, you'll see an opening."[11]

What does it take for that opening to be comprehensive climate policy, and when does one of the lines in the presidential decision memo include an entry for "solar geoengineering"?

Notes

Introduction: Start here – But don't start with geoengineering

1 See Broecker (1975).
2 See Revelle et al. (1965). This report is often billed as the first ever report to a president on climate change. In fact, John F. Kennedy, too, received a (brief) climate change warning, and so has every president since (Hulac, 2018).
3 Budyko's proposal first appeared in Russian (Budyko, 1974), subsequently translated into English (Budyko, 1977). See Caldeira and Bala (2017) for a brief history of the idea. Morton (2015) reviews the history in depth.
4 See National Research Council (1992).
5 See Crutzen (2006).
6 See Navarro et al. (2016).
7 See Cicerone (2006).
8 The clothing example is imperfect for another reason. The additional heat absorbed by black outerwear is typically lost before it reaches the skin. See Shkolnik et al.'s (1980) aptly named *Nature* study: "Why do Bedouins wear black robes in hot deserts?"
9 See e.g. Ocko et al. (2017).
10 See *The Economist* (2008).
11 See e.g. Goodell (2017).
12 Keith (2000) first mentions the three core characteristics. Keith, Parson, and Morgan (2010) first mentions the exact phrase: "fast, cheap, and imperfect." Parson

and Ernst (2013) explores its governance implications, Moreno-Cruz, Wagner, and Keith (2018) its formal economic implications, and Mahajan, Tingley, and Wagner (2019) U.S. public opinion of these characteristics.

13 See table 2 in Smith and Wagner (2018). Also see Smith (2020) as well as Lockley, MacMartin, and Hunt (2020).

14 See Gingrich (2008).

15 See Baker and Wagner (2016), and Moreno-Cruz, Wagner, and Keith (2018) for a formal exploration.

1 Not *if*, but *when*

1 See Sherwood et al. (2020).

2 See Wagner and Weitzman (2015).

3 See note 15 on page 11.

4 See Pigou (1920, p. 161).

5 See Aklin and Mildenberger (2020).

6 See Wagner and Weitzman (2012) for the first mention of "free driver," and Weitzman (2015) for a formalization of the idea.

7 See e.g. Schelling (1996); Barrett (2008); and Victor (2008).

8 See Zeckhauser (2006) for a standard classification in the investment context. Also see Chapter 2.

9 See Stockholm International Peace Research Institute (2018) as well as the discussion in Smith and Wagner (2018). A further 15 countries have military budgets over $3 billion, bringing the total number of countries to 50.

10 See McClellan et al. (2010) for the original consulting report, resulting in a subsequent peer-reviewed analysis (McClellan et al., 2012). See the discussion in the introduction of Smith and Wagner (2018) for further studies as well as Lockley, MacMartin, and Hunt (2020) for an updated review.

11 See Keith (2013, p. 6).

12 Prior to Pemco, Wake Smith was the COO of Atlas Air Worldwide Holdings. He was also President of the training division of Boeing.

13 See Smith and Wagner (2018). See Buck and Wagner (2018) for some of the immediate media fallout, discussed further in "Too fast and/or too slow?" in Chapter 3.

Smith has since taken things further, calculating "mature" deployment costs (Smith, 2020) and joining forces with aircraft engineers to explore the SAIL aircraft design further (Bingaman et al., 2020).

14 Jacobsen et al. (2019) put the global figure at $73 trillion across 143 countries.

15 See Barrett's (2008) "The incredible economics of geoengineering" for an astute early take.

16 See, for example, Horton and Reynolds (2016) and also Reynolds and Wagner (2019).

17 See note 3 on page 3.

18 See Crutzen (2006). See Necheles, Burns, and Keith (2018) for a detailed look at solar geoengineering research funding.

19 See U.S. Global Change Research Program (2019).

20 See note 10 on page 18.

21 See Victor (2008, p. 324) as well as Wagner and Weitzman (2015, chap. 6) for an attempt at just such a screenplay.

22 I counseled Jeff Bezos not to "tilt advocacy much too quickly into one direction" by spending even a fraction of his pledge on solar geoengineering research, let alone deployment: "There is similar danger in distorting the overall climate policy and technology landscape with one person's preferences. Imagine a (crazed) billionaire spending $1 billion on deploying solar geoengineering now" (Wagner, 2020a). *The Economist* (2020), meanwhile, was more direct, calling for a doubling of current solar geoengineering funding of $20 million annually.

23 See Bodansky (2013, p. 548).

24 See Reynolds and Wagner (2019) as well as Chapter 6.

25 See Roth (1993) for the early history of experimental economics, including a description of the game conducted by Melvin Dresher and Merrill Flood in 1950 that would later be named the "prisoner's dilemma." Poundstone (1993) presents this and many more such dilemmas in game theory.

26 See, for example, Thomson (1985) and Edmonds (2013).

27 See the discussion around table 1 in Fabre and Wagner (2020) for more details.

28 As I argue with Adrien Fabre in Fabre and Wagner (2020), while others model total mitigation as the sum of each player's contribution (Barrett, 2007; Sandler, 2018), this may not just apply to climate negotiations but also to mitigation efforts *per se*: "If the combination of diverging mitigation efforts ([L_1, H_2] or [H_1, L_2]) led to moderate overall mitigation effort M, it might be reasonable to treat M and L the same for purposes of this analysis. That might hold especially if there is a threshold of sorts between M and H. If so, only the outcome H might lead to stabilizing global climates, while both L and M would imply indefinite warming, albeit at different speeds."
29 See Barrett (2007).
30 See e.g. Gillingham and Stock (2018).
31 See Aklin and Mildenberger (2020).
32 See e.g. Kothen (2018) and Nordhaus (2015).
33 See e.g. Wagner (2020b).
34 See table 2 in Fabre and Wagner (2020) and the discussion around it for further details.
35 Table and description taken from Fabre and Wagner (2020), table 3.
36 See Fabre and Wagner (2020). The solution is strikingly simple, going back to a subgame-perfect Nash equilibrium in a non-cooperative game (Nash, 1951; Selten, 1965). Adrien Fabre and I are not the first to point to this possibility (Fabre and Wagner, 2020). Others have pointed to the possibility in the context of strategic climate negotiations (Millard-Ball, 2012; Moreno-Cruz, 2015; Urpelainen, 2012). At least one study looking to public opinions also points to the possibility of how mere mention of solar geoengineering might prompt some to want to mitigate more, in this case in the form of buying more carbon offsets (Merk et al., 2016). (See note 28 on page 130 and the discussion around it.)

2 What could possibly go wrong?

1 Yang, Peltier, and Hu (2012) show how Snowball Earth depends on the atmospheric concentration of CO_2. Pre-industrial CO_2 concentrations of around 286 ppm

require an 8–9% decrease in solar radiation for Snowball Earth to occur. Higher CO_2 concentrations would require greater solar dimming. Snowball Earth, in turn, ended due to a massive release of CO_2, causing global temperature to increase (Hoffman et al., 1998; Hoffman and Schrag, 2002).

2 See Robock (2008). While the title only says "geoengineering," Robock does indeed refer to solar geoengineering and within it primarily stratospheric aerosols. A later update cites 28 reasons, contrasted with six potential benefits (Robock, 2020).

3 One of the largest and most comprehensive stratospheric aerosol modeling efforts assumes injections at both 15° and 30° north and south (Tilmes et al., 2018).

4 Hemispheric or regional solar geoengineering could have substantial effects on rainfall pattern, for example in the Sahel (Haywood et al., 2013). It could also have unintended effects on tropical cyclones. Solar geoengineering just in the southern hemisphere could, for example, increase cyclone frequency (Jones et al., 2017).

5 See Mersereau (2016) and Wood (2014). For a book-length, factual treatment see Klingaman and Klingaman (2013). For a more literary account, see Glasfurd (2020).

6 See Proctor et al. (2018). More on that under point 4, "Effects on plants," and 20, "Unexpected consequences."

7 Peter Stott leads the Climate Monitoring and Attribution team of the Hadley Centre for Climate Prediction and Research at the U.K. Met Office. See Stott et al. (2013) for an overview of this work. See Stott et al. (2004) for an early example, focused on the European heatwave of 2003. See also Trenberth et al. (2015) for an independent view.

8 Robock (2008) here cites Trenberth and Dai (2007), who point to some possible links. See Robock (2000) for a more comprehensive review of volcanic effects on the climate.

9 See Oman et al. (2005), also referenced in Robock (2008), and Robock et al. (2008). Both Oman et al. (2005) and Robock et al. (2008) prominently mention possible

impacts on the monsoon in their abstract, something specifically pointed out by Keith (2013, p. 57) in the case of Robock et al. (2008).

10 See Robock et al. (2008).

11 See Keith (2013, pp. 57–58), who, in turn, cites Gupta (2010).

12 Keith (2013) says how a perfunctory Google search in April 2012 showed how, "More than half of all Google hits for 'geoengineering' now include 'monsoon'" (p. 57, and note 16). That is no longer the case, but Keith's broader point of the monsoon's importance surely stands.

13 See, for example, Climate Central (2017).

14 See, for example, Oreskes (2019) on the overall point on why to trust science in the first place.

15 For book-length introductions to Science, Technology, and Society aka Science and Technology Studies (both STS), see, for example, Jasanoff (2016, 2011, 2009, 2004) or also Jasanoff et al. (2001). And yes, my exclusively citing Sheila Jasanoff here, someone who has introduced me to STS in general and, in an undergraduate survey course, environmental politics more specifically, I, too, show my limited worldview and biases.

16 See Pongratz et al. (2012).

17 See Proctor et al. (2018), and points 4, "Effects on plants," and 20, "Unexpected consequences," below.

18 See, among many others, Rogelj et al. (2012) on representative concentration pathways. The Kaya identity dissects the chain from economic activity per capita into emissions in its various elements, including population, economic activity, energy intensity of production, and emissions intensity of energy generation (Commoner et al., 1971; Ehrlich and Holdren, 1971; Kaya, 1989). Successive IPCC assessment and special reports summarize many of these literatures, from IPCC (1992) through IPCC (2018, 2013). The Sixth Assessment Report is currently in the works.

19 See Irvine et al. (2019). The paper's other authors: MIT scientist and tropical cyclone expert Kerry Emanuel; Jie He, Larry Horowitz, and Gabriel Vecchi all having

worked with a model built at Princeton's Geophysical Fluid Dynamics Laboratory; and David Keith. (Also see note 22 on page 48.)

20 See Keith and Irvine (2016).

21 To be precise, the temperature increase now declines to +0.93°C, almost half of 2°C (Irvine et al., 2019).

22 See Irvine and Keith (2020), who confirm the prior results focusing on a halving of future warming with stratospheric aerosols, while relaxing the "idealized" part. The new paper employs a model of stratospheric aerosols rather than an "idealized' lowering of the solar constant.

23 The fifth metric, analyzed in the Geophysical Fluid Dynamics Lab model scenario run for the paper measures the power dissipation index of tropical cyclones, another well-established measure of cyclone strength (Emanuel, 2005). "Tropical cyclone" is the broad category. In the North Atlantic and Northeast Pacific – around the United States – they are typically called hurricanes. The same phenomenon in the Northwest Pacific is a typhoon. In the South Pacific and the Indian Ocean, it's a cyclone.

24 Irvine and Keith (2020) revisit the analysis pioneered by Irvine et al. (2019) and draw similar conclusions.

25 See Robock (2008, p. 15).

26 See National Research Council (2015a).

27 See Keith et al. (2017). I am one of the "et al." The other is Claire Zabel, who, as the extensive ethics declaration says, "began work on this analysis while a researcher at Harvard. She now works for the Open Philanthropy Project, which subsequently became a funder of Harvard's Solar Geoengineering Research Project, co-directed by [David Keith] and [me]." More on that in Chapter 3.

28 Kate Ricke's (2019) tweet summarizes the point. Cao and Jiang (2017) demonstrates it.

29 See e.g. Barrett (2003), Benedick (2009), and Parson (2003).

30 See Anderson et al. (1989).

31 See Keith et al. (2016).

32 See Dai et al. (2020).

33 In fall 2019, Jonathan Proctor began a post-doc at Harvard, joint between the Harvard Data Science

Initiative and the Harvard University Center for the Environment, in which Harvard's Solar Geoengineering Research Program is formally housed.

34 See Eastham et al. (2018).
35 See Kravitz et al. (2009).
36 See Lohmann and Gasparini (2017) for a survey. See Storelvmo et al. (2013) and also Muri et al. (2014) for early modeling work.
37 See Zerefos et al. (2007).
38 See Kolbert (2021).
39 See Proctor et al. (2018).
40 See e.g. Wagner (2020c, 2020d).
41 See Reynolds and Wagner (2019) as well as Chapter 6.
42 See Ocko et al. (2017). Kolbert (2014) puts some of the currently observed rates of change into historical context.
43 See Parker and Irvine (2018) as well as Reynolds et al. (2016) for a list of typical reasons – and for arguments about why termination shock is overplayed.
44 See Penna and Rivers (2013).
45 See Parker and Irvine (2018) and Reynolds et al. (2016).
46 See MacMartin et al. (2014a, 2014b).
47 See Chapter 1.
48 See Robock (2015).
49 See Keith and Dykema (2018).
50 See Reynolds (2019a) for a book-length treatise on governance, with a chapter on international law, and Gerrard and Hester (2018) for a comprehensive take on U.S. law.
51 In particular, watch the Carnegie Council for Ethics and International Affairs' Carnegie Climate Governance Initiative (C2G), previously known as Carnegie Climate Geoengineering Governance Initiative (C2G2), and its blog, www.c2g2.net/c2g-blog/.
52 For prominent papers addressing some of the questions here, see Szerszynski et al. (2013) for an argument on "why solar radiation management geoengineering and democracy won't mix" and Horton et al. (2018) for a good rejoinder. The latter counts as its co-authors both David Keith and the late Steve Rayner. Rayner himself has

written many more skeptical pieces, drawing attention to the many ethical questions (e.g., Heyward and Rayner, 2015). Also see note 18 on page 80 and the discussion around it in Chapter 3.

53 See Tingley and Wagner (2017).
54 See *The Economist* (2018)
55 See Shepherd (2018).
56 See Miller and Reynolds (2009).
57 See World Health Organization (2020).
58 See Shearer et al. (2016).
59 See *Popular Science* (1943).
60 See Wagner (2011a).
61 See Crutzen (2006).
62 See Wagner and Weitzman (2018).
63 See Chapter 7.
64 See, for example, Yaroshevsky (2006) and McLean (1976).
65 See Shearer et al. (2016).
66 See Wagner (2018).
67 See Robock (2008).
68 See Wagner and Weitzman (2015).
69 See Zeckhauser and Wagner (2019). I'm currently working on a model, jointly with Richard Zeckhauser and Maryaline Catillon, to document this possibility more formally.
70 See Visioni et al. (2021) for an exploration of how good a proxy 'turning down the sun' is.
71 See Sherwood et al. (2020).
72 See Kravitz and MacMartin (2020) for a review of solar geoengineering and specifically stratospheric aerosol uncertainties, and what science does, in fact, know. Kravitz and MacMartin also lay out important questions for further research.

3 The drive to research

1 See Revelle et al. (1965) as well as note 2 on page 3 and the brief discussion around it.
2 See Dales (1968). See Weitzman (1974) for the seminal paper comparing prices versus quantities.
3 See Moreno-Cruz et al. (2018).

4 It also corresponds to what's often called the "napkin" diagram, so called because it was first drawn on the proverbial napkin (more in Chapter 4).
5 See note 5 on page 75 and the text around it.
6 See Keith (2000). Also see Barrett (2008) and Victor (2008), and the discussion in Chapter 1.
7 See note 3 on page 3.
8 See National Research Council (1992, chap. 28).
9 See Mautner and Parks (1990). See Angel (2006) for a later set of calculations confirming the general feasibility.
10 See Keith and Dowlatabadi (1992).
11 See Teller (1968) as well as later discussions in Chapter 7 on the one hand, and Teller et al. (2002, 1999, 1997) on the other.
12 See Goodell (2011) and Caldeira and Bala (2017).
13 See Keith (2000).
14 See Lane et al. (2007)
15 See Keith (2000).
16 See MacCracken et al. (2010)
17 See Crutzen (2006) and the discussion around note 5 on page 4 in the text.
18 See Burns et al. (2019) for a publicly accessible Zotero database of scientific solar geoengineering publications, a project I was involved in while at Harvard. Also see note 52 on page 66 and the discussion around it.
19 The project website was first posted in early 2018. Its description has been adjusted since then, including a substantive Q&A (Keutsch, 2020). The page itself is on Frank Keutsch's lab page, largely maintained by the project team involving Keutsch (formally the Principal Investigator), David Keith, John Dykema, and Lizzie Burns – all of whom were close colleagues while I was working as founding co-director of Harvard's Solar Geoengineering Research Program. Lizzie Burns took over my executive director job as managing director of the Program. I had never been part of the SCoPEx project team myself.
20 See Pierrehumbert (2017) and Pierrehumbert (2015), respectively. The former is a blog entry on the site of the *Bulletin of the Atomic Scientists*, which later published a

peer-reviewed report as well (Pierrehumbert, 2019a). The organization had previously published Alan Robock's "20 reasons why geoengineering may be a bad idea" (Robock, 2008). (See Chapter 2.)

21 See Pierrehumbert (2019b).

22 See, for example, Keith and Wagner (2017) as a response to one such misguided headline.

23 See Dykema et al. (2014). John Dykema, the paper's first author, is both a member of the Keith and Keutsch Groups at Harvard, and the project scientist on SCoPEx.

24 The early separation of SCoPEx from the broader Solar Geoengineering Research Program (SGRP) was real. SCoPEx applied for funding from SGRP much like other faculty projects. By now the roles of managing SGRP and SCoPEx are indeed intertwined, though only for a very practical reason. Neither role commands a fulltime position.

25 See Keith and Dykema (2018), and the discussion in section "15 and 16. Commercial and military control of the technology" of Chapter 2.

26 Harvard's Solar Geoengineering Research Program defines it as such: "Geoengineering is conventionally split into two broad categories: The first is carbon geoengineering, often also called carbon dioxide removal (CDR). The other is Solar Geoengineering, often also called Solar Radiation Management (SRM), albedo modification, or sunlight reflection. There are large differences" (Harvard's Solar Geoengineering Research Program, 2016). I wrote those words.

27 See www.scopexac.com.

28 See National Academies of Sciences, Engineering, and Medicine (2021).

29 See note 18 on page 80 and the discussion around it.

30 See Kravitz and MacMartin (2020).

31 See the discussion in Chapter 2, especially descriptions of results from Irvine et al. (2019) and also Irvine and Keith (2020).

32 See Smith and Wagner (2018).

33 See McClellan et al. (2012). *Environmental Research*

Letters has since published a further solar stratospheric aerosol costing study (Smith, 2020).

34 See Carrington (2018).
35 See Davenport and Pierre-Louis (2018).
36 See Buck and Wagner (2018) for a description of some of the fallout. See Keck (2018) for a news story putting the paper and subsequent media fallout into some context.
37 See Necheles et al. (2018) and the discussion around note 18 on page 22.
38 See Rahman et al. (2018).
39 See Pinto et al. (2020), a team based at the University of Cape Town, South Africa, with one U.S.-based collaborator: Simone Tilmes from the National Center for Atmospheric Research in Boulder, CO. DECIMALS, in turn, is run by SRM Governance Initiative, which is directed by Andrew Parker, who had previously been a researcher with David Keith at Harvard.

A warning

1 See Climate Files (1998).

4 "Rational" climate policy

1 See Wagner (2020e) for a non-fictional call for the importance of systemic change to address systemic risks.
2 See Long and Shepherd (2014) for a broader discussion. John Shepherd first presented the original "napkin diagram" at the March 2010 Asilomar International Conference on Climate Intervention Technologies. This version is one I have used as a rough sketch to explain the various interventions, based on a similar version used frequently by David Keith.
3 See note 9 on page 18 and the discussion around it.
4 See Keith and Irvine (2016) for suggesting this half–half split as a basis for further research. See Irvine et al. (2019) as well as Irvine and Keith (2020) for modeling research doing just that. Others have begun to take up this research hypothesis as well. I, for example, did so here as the basis of a stratospheric aerosol costing scenario: Smith and Wagner (2018).

5 See U.K. Royal Society (2009).
6 See National Research Council (1992).
7 See National Research Council (2015a), which focuses on solar geoengineering, while National Research Council (2015b) focuses on carbon removal.
8 See National Academies of Sciences, Engineering, and Medicine (2021).
9 See www.c2g2.net/geoeng-sdgs. (C2G's original name included a second 'G' for 'Geoengineering', since dropped, hence the URL including a second '2'.)
10 See www.c2g2.net/c2g2-mission and also www.c2g2.net/our-approach with more details on C2G's approach.

5 A humanitarian cyclone crisis

1 See, for example, Mufson et al. (2019).
2 See Xu et al. (2020).
3 Box TS.5, figure 1, reproduced with permission. See IPCC (2007b). Other versions of this same schematic show how increases in the mean combined with increases in the variance in global average temperatures can lead to even greater increases in extreme weather events (e.g., IPCC, 2001, fig. 2.32, Working Group I).
4 See McDonald et al. (2019).
5 See Smith and Wagner (2018) for some research supporting what, it must be emphasized, is indeed an entirely hypothetical scenario. No, Embraer, to my knowledge, is not, in fact, working on this – and no, Dear Conspiracy Theorists, the plane above your head, Embraer or otherwise, is not currently geoengineering the planet. If you have any doubt on that latter point, see "18 and 19. Control of the thermostat and questions of moral authority" in Chapter 2.
6 See discussion of SCoPEx in Chapter 3. Also see Keith et al. (2016) for calcium carbonate.

6 Millions of geoengineers

1 See Smith and Wagner (2018), Smith (2020), as well as notes 12 and 13 on pages 19 and 20, respectively, and the discussion around them.
2 See Weaver (1986).

3 See Ricke et al. (2018), in particular figure 4, with Saudi Arabia having the highest share of the Global Social Cost of Carbon after India and on a par with the United States.
4 See Gillingham et al. (2016) on the rebound effect, Dechezleprêtre and Sato (2017) on spatial leakage, and Jensen et al. (2015) on temporal leakage, aka the "Green Paradox."
5 See note 21 on page 24 and the discussion around it.
6 Reproduced with permission from Reynolds and Wagner (2019). See Lockley, MacMartin, and Hunt (2020) for a recent review of alternative deployment methods.
7 See the World Drug Report, published annually by the UN Office on Drugs and Crime, available at: wdr.unodc.org.
8 See Reynolds and Wagner (2019) for a more in-depth discussion and Lockley, MacMartin, and Hunt (2020) for the latest review.
9 See e.g. Hofstadter (1964) or Machiavelli (1532).
10 See Hume (1740).

7 Green moral hazards

1 This chapter is substantively based on Wagner and Zizzamia (2021), which we briefly summarized for a popular audience in Wagner and Zizzamia (2020). The core argument also appears in Wagner and Merk (2018), in turn summarized in Wagner and Merk (2019). For other comprehensive takes, see Reynolds (2019a, chap. 3) and Lin (2013).
2 See Finkelstein (2014).
3 See Gillingham et al. (2016) as well as the discussion in and around note 177 in the prior chapter.
4 See Keith (2000), who first introduced the term in the context of geoengineering.
5 See McLaren (2016) and Morrow (2014), respectively. Also see Lin (2013) for a comprehensive discussion and Reynolds (2015) for a critical review.
6 See Gingrich (2008).
7 See Wagner (2012) for some of my biases in that regard,

a set of arguments I reiterate more recently, joint with Martin Weitzman, in the final chapter of *Climate Shock* (Wagner and Weitzman, 2015).

8 See e.g. Wagner (2020f) for a recent summary of the main argument. See e.g. Jenkins et al. (2020) for more in-depth discussions.

9 See e.g. Wagner et al. (2015) and Meckling, Sterner, and Wagner (2017).

10 See note 10 on page 8 and the discussion around it.

11 See Wagner and Zizzamia (2020, 2021).

12 See e.g. Mann (2018).

13 I told both myself, in chapter 6 of *But Will the Planet Notice?* (Wagner, 2011a).

14 See Meyer (2010) and, more recently, Ridley (2015), also described in Wagner (2011a, chap. 6).

15 See Bardi (2004), largely based on data by Starbuck (1878), and later described in more detail in Bardi (2011). Bardi and Lavacchi (2009) show detailed whale-oil data. Note that the label in their figure 5 should read "year 0 = 1818," not "1804," a typo I confirmed with the authors while writing Wagner (2011a, chap. 6). See Brox (2010) for another eloquent retelling.

16 The horse-manure story has been told many times. This excerpt here is from chapter 6 of *But Will the Planet Notice?* (Wagner, 2011a), which, in turn, takes some facts from Morris (2007). For a terrific take on the horse-manure story in the context of solar geoengineering, see Kolbert (2009), who also clears up some misconceptions propagated by Levitt and Dubner's (2011) *Superfreakonomics*.

17 See Wagner (2011a, chap. 6), which, in turn, is based on Flink (1990) and Yago (1984).

18 See Jessee (2013).

19 See Commoner (1966, p. 64).

20 See Carson (1962).

21 See Revelle et al. (1965).

22 See Borgman (2012) and Morton (2015, pp. 153–154, 157).

23 See Brand (1968).

24 See Brand (2009).
25 See Burns et al. (2016). That review focuses on solar geoengineering. Some studies also include carbon removal explicitly. There are some other studies, not captured here, which solely focus on carbon removal technologies.
26 See Podsakoff et al. (2003). In Mahajan, Tingley, and Wagner (2019), we explicitly test for acquiescence bias around moral hazard in solar geoengineering and confirm that it dwarfs otherwise weak moral hazard findings.
27 My own bias on the offset question: voluntary ones are largely pointless and might even imply a step backward (Wagner, 2011a, 2011b).
28 See Merk, Pönitzsch, and Rehdanz (2016).
29 See Wagner and Merk (2019, 2018). Watch out for "Merk and Wagner" coming out hopefully soon-ish, summarizing our further research. See Maki et al. (2019) for a survey of the broader literature on pro-environment behavioral spillover related to the 'inverse moral hazard' phenomenon.
30 See Wagner and Weitzman (2015), and many an essay before and since (e.g., Wagner and Weitzman, 2018, 2012).
31 See Holthaus (2020).
32 See note 1 on page 35 and the discussion around it.
33 See Peper (2020) and Robinson (2020).
34 See www.scopexac.com as well as discussions in Chapter 3.
35 See Frumhoff and Stephens (2018).

8 Research governance

1 See Gerrard and Hester (2018) for the first book-length treatment exclusively focused on geoengineering and the law; Reynolds (2019a) for the most comprehensive treatment focused on solar geoengineering; and Pasztor, Scharf, and Schmidt (2017) for a brief summary of the main issues.
2 See the U.K. Royal Society (2009) for the first comprehensive such solar geoengineering report. The first U.S. report with an explicit mention of solar geoengineering

was published in 1992 (National Research Council, 1992), followed by National Research Council (2015a), while National Research Council (2015b) focuses on carbon removal. Lastly, see National Academies of Sciences, Engineering, and Medicine (2021).

3 See Temple (2019).
4 See Necheles, Burns, and Keith (2018).
5 See National Academies of Sciences, Engineering, and Medicine (2021).
6 See Parson and Keith (2013).
7 See Parson and Keith (2013).
8 See Cicerone (2006) and Crutzen (2006).
9 See MacCracken et al. (2010) and the discussion of Asilomar in Chapter 3.
10 See Parson and Keith (2013).
11 For one well-documented example of the rather disingenuous nature of some of the workings of the ETC Group, see Reynolds (2019b).
12 See Pasztor, Scharf, and Schmidt (2017).
13 See www.cbd.int/decisions/cop/13/14.
14 See, for example, Wagner and Weitzman (2015).
15 See Chapter 7.
16 See "What if geoengineering could lead to a more ambitious mitigation agreement?" in Chapter 1.
17 See Robock (2015) and section "15 and 16. Commercial and military control of the technology" in Chapter 2.
18 See Low and Buck (2020).

Epilogue: The inevitable gamble

1 This statement goes back to an unfortunate headline in the *Guardian* (Watts, 2018), whose own environment editor, on leave at the time, says he would not have framed it that way. Costa Samaras and I discuss its implications (Wagner and Samaras, 2019).
2 See Wagner and Weitzman (2015), itself grown out of a *Foreign Policy* essay squarely focused on geoengineering (Wagner and Weitzman, 2012).
3 See Shabecoff (1988). Many a lore has persisted for decades about that pivotal hearing, including that the

hearing room was intentionally made hotter for dramatic effect. It was not (Kessler, 2015).

4 See Banerjee, Song, and Hasemyer (2015).
5 See Revelle et al. (1965).
6 See Arrhenius (1896).
7 See Molena (1912).
8 See e.g. Wagner (2020g).
9 See Proctor et al. (2018) and "4. Effects on plants" in Chapter 2.
10 See Merk, Pönitzsch, and Rehdanz (2016) and "Moral hazard and its inverse?" in Chapter 7.
11 See https://www.youtube.com/watch?v=UM-Q_zpuJGU&t=775s.

References

Aklin, M., Mildenberger, M., 2020. Prisoners of the wrong dilemma: Why distributive conflict, not collective action, characterizes the politics of climate change. *Global Environmental Politics* 20, 4–27. https://doi.org/10.1162/glep_a_00578

Anderson, J.G., Brune, W.H., Proffitt, M.H., 1989. Ozone destruction by chlorine radicals within the Antarctic vortex: The spatial and temporal evolution of ClO-O3 anticorrelation based on in situ ER-2 data. *Journal of Geophysical Research: Atmospheres* 94, 11465–11479. https://doi.org/10.1029/JD094iD09p11465

Angel, R., 2006. Feasibility of cooling the Earth with a cloud of small spacecraft near the inner Lagrange point (L1). *Proceedings of the National Academy of Sciences* 103, 17184–17189.

Arrhenius, S., 1896. On the influence of carbonic acid in the air upon the temperature of the ground. *London, Edinburgh, and Dublin Philosophical Magazine and Journal of Science* 41, 237–276.

Baker Jr., G.L., Wagner, G., 2016. Need portfolio approach to climate risk. *The Mercury News/East Bay Times*.

Banerjee, N., Song, L., Hasemyer, D., 2015. Exxon: The road not taken [WWW Document]. *InsideClimate News*. https://insideclimatenews.org/news/15092015/Exxons-own-research-confirmed-fossil-fuels-role-in-global-warming

Bardi, U., 2011. *The Limits to Growth Revisited*. Springer Science and Business Media.

Bardi, U., 2004. Prices and production over a complete Hubbert cycle: The case of the American whaling industry in the 19th century. *Association for the Study of Peak Oil and Gas.*

Bardi, U., Lavacchi, A., 2009. A simple interpretation of Hubbert's model of resource exploitation. *Energies* 2, 646–661.

Barrett, S., 2008. The incredible economics of geoengineering. environ resource. *Econ* 39, 45–54. https://doi.org/10.1007/s10640-007-9174-8

Barrett, S., 2007. *Why Cooperate?: The Incentive to Supply Global Public Goods*. Oxford University Press on Demand.

Barrett, S., 2003. *Environment and Statecraft: The Strategy of Environmental Treaty-Making*. Oxford University Press.

Benedick, R.E., 2009. *Ozone Diplomacy: New Directions in Safeguarding the Planet*. Harvard University Press.

Bingaman, D.C., Rice, C.V., Smith, W., Vogel, P., 2020. A stratospheric aerosol injection lofter aircraft concept: Brimstone Angel, in AIAA Scitech 2020 Forum, AIAA SciTech Forum. *American Institute of Aeronautics and Astronautics*. https://doi.org/10.2514/6.2020-0618

Bodansky, D., 2013. The who, what, and wherefore of geoengineering governance. *Climatic Change* 121, 539–551.

Borgmann, A., 2012. The setting of the scene:

technological fixes and the design of the good life, in Preston, C.J. (ed.), *Engineering the Climate: The Ethics of Solar Radiation Management*. Lexington Books, Lanham, MD, pp. 189–200.

Brand, S., 2009. *Whole Earth Discipline: Why Dense Cities, Nuclear Power, Transgenic Crops, Restored Wildlands, and Geoengineering are Necessary*. Viking Press, New York, N.Y.

Brand, S., 1968. *The Whole Earth Catalog*. Portola Institute.

Broecker, W.S., 1975. Climatic change: Are we on the brink of a pronounced global warming? *Science* 189, 460–463.

Brox, J., 2010. *Brilliant: The Evolution of Artificial Light*. Houghton Mifflin Harcourt.

Buck, H., Wagner, G., 2018. What it's like when Questlove does a better job tweeting about your research than CNN. Harvard's Solar Geoengineering Research blog. https://gwagner.com/questlove-erl-cnn/

Budyko, M.I., 1977. Climatic changes. *American Geophysical Union*.

Budyko, M.I., 1974. Izmenenie klimata. *Gidrometeoizdat*.

Burns, E.T., Flegal, J.A., Keith, D.W., Mahajan, A., Tingley, D., Wagner, G., 2016. What do people think when they think about solar geoengineering? A review of empirical social science literature, and prospects for future research. *Earth's Future* 4, 536–542.

Burns, L., Chang, A., Irvine, P.J., Matzner, N., Necheles, E., Reynolds, J.L., Wagner, G., 2019. Solar Geoengineering Research Zotero Library [WWW Document]. https://geoengineering.environment.harvard.edu/blog/zotero

Caldeira, K., Bala, G., 2017. Reflecting on 50 years of geoengineering research. *Earth's Future* 5, 10–17. https://doi.org/10.1002/2016EF000454

Cao, L., Jiang, J., 2017. Simulated effect of carbon cycle feedback on climate response to solar geoengineering. *Geophysical Research Letters* 44, 12,484–12,491. https://doi.org/10.1002/2017GL076546

Carrington, D., 2018. Solar geoengineering could be "remarkably inexpensive" – report. *Guardian.*

Carson, R., 1962. *Silent Spring.* Houghton Mifflin Harcourt, New York, N.Y.

Catie Keck, 2018. No, scientists didn't just suggest we "dim the sun" to stop climate change. *Earther.*

Cicerone, R.J., 2006. Geoengineering: Encouraging research and overseeing implementation. *Climatic Change* 77, 221–226.

Climate Central, 2017. Small Change in Average, Big Change in Extremes [WWW Document]. https://www.climatecentral.org/gallery/graphics/small-change-in-average-big-change-in-extremes

Climate Files, 1998. 1998 Shell Internal TINA Group Scenarios 1998–2020 Report. http://www.climatefiles.com/shell/1998-shell-internal-tina-group-scenarios-1998-2020-report/

Commoner, B., 1966. *Science and Survival.* Viking Press, New York, N.Y.

Commoner, B., Corr, M., Stamler, P.J., 1971. The causes of pollution. *Environment: Science and Policy for Sustainable Development* 13, 2–19.

Crutzen, P.J., 2006. Albedo enhancement by stratospheric sulfur injections: A contribution to resolve a policy dilemma? *Climatic Change* 77, 211–220. https://doi.org/10.1007/s10584-006-9101-y

Dai, Z., Weisenstein, D.K., Keutsch, F.N., Keith, D.W., 2020. Experimental reaction rates constrain estimates of ozone response to calcium carbonate geoengineering. *Communications Earth and Environment* 1, 1–9. https://doi.org/10.1038/s43247-020-00058-7

Dales, J.H., 1968. *Pollution, Property, and Prices: An*

Essay in Policy-making and Economics. University of Toronto Press.
Davenport, C., Pierre-Louis, K., 2018. U.S. climate report warns of damaged environment and shrinking economy. *New York Times*.
Dechezleprêtre, A., Sato, M., 2017. The impacts of environmental regulations on competitiveness. *Review of Environmental Economics and Policy* 11, 183–206.
Dykema, J.A., Keith, D.W., Anderson, J.G., Weisenstein, D., 2014. Stratospheric controlled perturbation experiment: a small-scale experiment to improve understanding of the risks of solar geoengineering. *Philosophical Transactions of the Royal Society A: Mathematical, Physical and Engineering Sciences* 372, 20140059. https://doi.org/10.1098/rsta.2014.0059
Eastham, S.D., Weisenstein, D.K., Keith, D.W., Barrett, S.R.H., 2018. Quantifying the impact of sulfate geoengineering on mortality from air quality and UV-B exposure. *Atmospheric Environment* 187, 424–434. https://doi.org/10.1016/j.atmosenv.2018.05.047
Economist, 2020. Jeff Bezos wants to help save the climate. Here is how he should do it. *The Economist*.
Economist, 2018. Could Tibetan clouds save China from drought? *The Economist*.
Economist, 2008. Adapt or die. *The Economist*.
Edmonds, D., 2013. *Would You Kill the Fat Man?: The Trolley Problem and what Your Answer Tells Us About Right and Wrong*. Princeton University Press.
Ehrlich, P.R., Holdren, J.P., 1971. Impact of population growth. *Science* 171, 1212–1217.
Emanuel, K., 2005. Increasing destructiveness of tropical cyclones over the past 30 years. *Nature* 436, 686–688. https://doi.org/10.1038/nature03906
Fabre, A., Wagner, G., 2020. *Availability of Risky Geoengineering Can Make an Ambitious Climate*

Mitigation Agreement More Likely. Palgrave Communications.
Finkelstein, A., 2014. *Moral Hazard in Health Insurance*. Columbia University Press.
Flink, J.J., 1990. *The Automobile Age*. MIT Press.
Frumhoff, P.C., Stephens, J.C., 2018. Towards legitimacy of the solar geoengineering research enterprise. *Philosophical Transactions of the Royal Society A: Mathematical, Physical and Engineering Sciences* 376, 20160459.
Gerrard, M.B., Hester, T., 2018. *Climate Engineering and the Law: Regulation and Liability for Solar Radiation Management and Carbon Dioxide Removal*. Cambridge University Press.
Gillingham, K., Rapson, D., Wagner, G., 2016. The rebound effect and energy efficiency policy. *Review of Environmental Economics and Policy* 10, 68–88.
Gillingham, K., Stock, J.H., 2018. The cost of reducing greenhouse gas emissions. *Journal of Economic Perspectives* 32, 53–72. https://doi.org/10.1257/jep.32.4.53
Gingrich, N., 2008. Stop the green pig: Defeat the Boxer-Warner-Lieberman Green Pork Bill capping American jobs and trading America's future. *Human Events*.
Glasfurd, G., 2020. *The Year Without Summer: 1816 – One Event, Six Lives, a World Changed*. John Murray Press.
Goodell, J., 2017. *The Water Will Come: Rising Seas, Sinking Cities, and the Remaking of the Civilized World*. Little, Brown.
Goodell, J., 2011. Can geoengineering save the world? *Rolling Stone*. https://www.rollingstone.com/politics/politics-news/can-geoengineering-save-the-world-238326/
Gupta, A., 2010. Geoengineering the planet? *Z Magazine*.

Harvard's Solar Geoengineering research program, 2016. *Geoengineering* [WWW Document]. https://geoengineering.environment.harvard.edu/geoengineering

Haywood, J.M., Jones, A., Bellouin, N., Stephenson, D., 2013. Asymmetric forcing from stratospheric aerosols impacts Sahelian rainfall. *Nature Climate Change* 3, 660–665.

Heyward, C., Rayner, S., 2015. Uneasy expertise: Geoengineering, social science, and democracy in the Anthropocene. Oxford University.

Hoffman, P.F., Kaufman, A.J., Halverson, G.P., Schrag, D.P., 1998. A Neoproterozoic Snowball Earth. *Science* 281, 1342–1346. https://doi.org/10.1126/science.281.5381.1342

Hoffman, P.F., Schrag, D.P., 2002. The Snowball Earth hypothesis: Testing the limits of global change. *Terra Nova* 14, 129–155. https://doi.org/10.1046/j.1365-3121.2002.00408.x

Hofstadter, R., 1964. *The Paranoid Style in American Politics*. Vintage.

Holthaus, E., 2020. *The Future Earth: A Radical Vision for What's Possible in the Age of Warming*. HarperOne, New York, N.Y.

Horton, J.B., Reynolds, J.L., 2016. The international politics of climate engineering: A review and prospectus for international relations. *International Studies Review* 18, 438–461. https://doi.org/10.1093/isr/viv013

Horton, J.B., Reynolds, J.L., Buck, H.J., Callies, D., Schäfer, S., Keith, D.W., Rayner, S., 2018. Solar geoengineering and democracy. *Global Environmental Politics* 18, 5–24.

Hulac, B., 2018. Every president since JFK was warned about climate change. *ClimateWire*.

Hume, D., 1740. *A Treatise of Human Nature*. London.

IPCC, 2018. Global warming of 1.5°C.

IPCC, 2013. Fifth Assessment Report: Climate Change.

IPCC, 2007a. Fourth Assessment Report.

IPCC, 2007b. Technical Summary. In *Climate Change 2007: The Physical Science Basis. Contribution of Working Group I to the Fourth Assessment Report of the Intergovernmental Panel on Climate Change* [Solomon, S., D. Qin, M. Manning, Z. Chen, M. Marquis, K.B. Averyt, M. Tignor and H.L. Miller (eds)]. Cambridge University Press, Cambridge, U.K. and New York, N.Y., U.S.A.

IPCC, 2001. Third Assessment Report.

IPCC, 1992. First Assessment Report.

Irvine, P., Emanuel, K., He, J., Horowitz, L.W., Vecchi, G., Keith, D., 2019. Halving warming with idealized solar geoengineering moderates key climate hazards. *Nature Climate Change* 9, 295–299. https://doi.org/10.1038/s41558-019-0398-8

Irvine, P.J., Keith, D.W., 2020. Halving warming with stratospheric aerosol geoengineering moderates policy-relevant climate hazards. *Environmental Research Letters* 15, 044011. https://doi.org/10.1088/1748-9326/ab76de

Jacobson, M.Z., Delucchi, M.A., Cameron, M.A., Coughlin, S.J., Hay, C.A., Manogaran, I.P., Shu, Y., Krauland, A.-K. von, 2019. Impacts of green new deal energy plans on grid stability, costs, jobs, health, and climate in 143 countries. *One Earth* 1, 449–463. https://doi.org/10.1016/j.oneear.2019.12.003

Jasanoff, S., 2016. *The Ethics of Invention: Technology and the Human Future*. W.W. Norton & Company.

Jasanoff, S., 2011. *Designs on Nature: Science and Democracy in Europe and the United States*. Princeton University Press.

Jasanoff, S., 2009. *The Fifth Branch: Science Advisers as Policymakers*. Harvard University Press.

Jasanoff, S., 2004. *States of Knowledge: The*

Co-Production of Science and the Social Order. Routledge.

Jasanoff, S., Markle, G.E., Peterson, J.C., Pinch, T., 2001. *Handbook of Science and Technology Studies*. Sage Publications.

Jenkins, J., Stokes, L., Wagner, G., 2020. *Carbon Pricing and Innovation in a World of Political Constraints* (Workshop Report). N.Y.U. Wagner, New York, N.Y.

Jensen, S., Mohlin, K., Pittel, K., Sterner, T., 2015. An introduction to the green paradox: The unintended consequences of climate policies. *Review of Environmental Economics and Policy* 9, 246–265.

Jessee, E.J., 2013. Radiation ecologies: Bombs, bodies, and environment during the atmospheric nuclear weapons testing period, 1942–1965 (PhD Thesis). Montana State University-Bozeman, College of Letters and Science.

Jones, A.C., Haywood, J.M., Dunstone, N., Emanuel, K., Hawcroft, M.K., Hodges, K.I., Jones, A., 2017. Impacts of hemispheric solar geoengineering on tropical cyclone frequency. *Nature Communications* 8, 1–10. https://doi.org/10.1038/s41467-017-01606-0

Kaya, Y., 1989. Impact of carbon dioxide emission control on GNP growth: interpretation of proposed scenarios. Intergovernmental Panel on Climate Change/Response Strategies Working Group, May.

Keith, D., 2013. *A Case for Climate Engineering*. The MIT Press, Cambridge, MA.

Keith, D., Dowlatabadi, H., 1992. A serious look at geoengineering. *Eos, Transactions American Geophysical Union* 73, 289–293.

Keith, D.W., 2000. Geoengineering the climate: History and prospect. *Annual Review of Energy and the Environment* 25, 245–284.

Keith, D.W., Dykema, J.A., 2018. Why we chose not to patent solar geoengineering technologies.

https://keith.seas.harvard.edu/blog/why-we-chose-not-patent-solar-geoengineering-technologies

Keith, D.W., Irvine, P.J., 2016. Solar geoengineering could substantially reduce climate risks – A research hypothesis for the next decade. *Earth's Future* 4, 2016EF000465. https://doi.org/10.1002/2016EF000465

Keith, D.W., Parson, E., Morgan, M.G., 2010. Research on global sun block needed now. *Nature* 463, 426–427. https://doi.org/10.1038/463426a

Keith, D.W., Wagner, G., 2017. Fear of solar geoengineering is healthy – but don't distort our research. *Guardian.*

Keith, D.W., Wagner, G., Zabel, C.L., 2017. Solar geoengineering reduces atmospheric carbon burden. *Nature Climate Change* 7, 617–619.

Keith, D.W., Weisenstein, D.K., Dykema, J.A., Keutsch, F.N., 2016. Stratospheric solar geoengineering without ozone loss. *PNAS* 113, 14910–14914. https://doi.org/10.1073/pnas.1615572113

Kessler, G., 2015. Setting the record straight: The real story of a pivotal climate-change hearing. *Washington Post.*

Keutsch, F.N., 2020. SCoPEx [WWW Document]. https://projects.iq.harvard.edu/keutschgroup/scopex

Klingaman, W.K., Klingaman, N.P., 2013. *The Year Without Summer: 1816 and the Volcano That Darkened the World and Changed History.* Macmillan.

Kolbert, E., 2021. *Under a White Sky.* Random House LCC U.S.

Kolbert, E., 2014. *The Sixth Extinction: An Unnatural History.* Henry Holt & Company.

Kolbert, E., 2009. Hosed. *The New Yorker.*

Kotchen, M.J., 2018. Which social cost of carbon? A theoretical perspective. *Journal of the Association of Environmental and Resource Economists* 5, 673–694.

Kravitz, B., MacMartin, D.G., 2020. Uncertainty and the basis for confidence in solar geoengineering research. *Nature Reviews Earth and Environment* 1, 64–75. https://doi.org/10.1038/s43017-019-0004-7

Kravitz, B., Robock, A., Oman, L., Stenchikov, G., Marquardt, A.B., 2009. Sulfuric acid deposition from stratospheric geoengineering with sulfate aerosols. *Journal of Geophysical Research: Atmospheres* 114. https://doi.org/10.1029/2009JD011918

Lane, L., Caldeira, K., Chatfield, R., Langhoff, S., 2007. Workshop Report on Managing Solar Radiation (No. NASA/CP-2007-214558, A-070010). NASA.

Levitt, S.D., Dubner, S.J., 2011. *Superfreakonomics*. Sperling & Kupfer.

Lin, A., 2013. Does geoengineering present a moral hazard? *Ecology Law Quarterly* 40, 673–712.

Lockley, A., MacMartin, D., Hunt, H., 2020. An update on engineering issues concerning stratospheric aerosol injection for geoengineering. *Environmental Research Commununications* 2, 082001. https://doi.org/10.1088/2515-7620/aba944

Lohmann, U., Gasparini, B., 2017. A cirrus cloud climate dial? *Science* 357, 248–249. https://doi.org/10.1126/science.aan3325

Long, J.C.S., Shepherd, J.G., 2014. The strategic value of geoengineering research, in Freedman, B. (ed.), *Global Environmental Change, Handbook of Global Environmental Pollution*. Springer Netherlands, pp. 757–770. https://doi.org/10.1007/978-94-007-5784-4_24

Low, S., Buck, H.J., 2020. The practice of responsible research and innovation in "climate engineering." *WIREs Climate Change* 11, e644. https://doi.org/10.1002/wcc.644

MacCracken, M., Berg, P., Crutzen, P.J., Barrett, S., Barry, R., Hamburg, S., Lampitt, R., Liverman, D., Lovejoy, T., McBean, G., Shepherd, J., Siedel, S., Somerville, R., Wigley, T., 2010. Statement from the Conference's Scientific Organizing Committee. Asilomar International Conference on Climate Intervention Technologies.

McClellan, J., Keith, D.W., Apt, J., 2012. Cost analysis of stratospheric albedo modification delivery systems. *Environmental Research Letters* 7, 034019.

McClellan, J., Sisco, J., Suarez, B., Keogh, G., 2010. *Geoengineering Cost Analysis*. Aurora Flight Sciences Corporation, Cambridge, MA.

McDonald, J., McGee, J., Brent, K., Burns, W., 2019. Governing geoengineering research for the Great Barrier Reef. *Climate Policy* 19, 801–811. https://doi.org/10.1080/14693062.2019.1592742

Machiavelli, N., 1532. *The Prince*. Translated by W. K. Marriott in 1908.

McLaren, D., 2016. Mitigation deterrence and the "moral hazard" of solar radiation management. *Earth's Future* 4, 596–602.

McLean, E.O., 1976. Chemistry of soil aluminum. *Communications in Soil Science and Plant Analysis* 7, 619–636. https://doi.org/10.1080/00103627609366672

MacMartin, D.G., Kravitz, B., Keith, D.W., Jarvis, A., 2014a. Dynamics of the coupled human–climate system resulting from closed-loop control of solar geoengineering. *Climate Dynamics* 43, 243–258. https://doi.org/10.1007/s00382-013-1822-9

MacMartin, D.G., Kravitz, B., Keith, D.W., 2014b. Geoengineering: The world's largest control problem. *American Control Conference*, 2401–2406.

Mahajan, A., Tingley, D., Wagner, G., 2019. Fast, cheap, and imperfect? U.S. public opinion about solar geoengineering. *Environmental Politics* 28, 523–543. https://doi.org/10.1080/09644016.2018.1479101

Maki, A., Carrico, A.R., Raimi, K.T., Truelove, H.B., Araujo, B., Yeung, K.L., 2019. Meta-analysis of pro-environmental behaviour spillover. *Nature Sustainability* 2, 307.

Mann, C.C., 2018. *The Wizard and the Prophet: Two Remarkable Scientists and Their Dueling Visions to Shape Tomorrow's World*. Knopf Doubleday Publishing Group.

Mautner, M., Parks, K., 1990. Space-based control of the climate, in *Engineering, Construction, and Operations in Space II*. ASCE, pp. 1159–1168.

Meckling, J., Sterner, T., Wagner, G., 2017. Policy sequencing toward decarbonization. *Nature Energy*. https://doi.org/10.1038/s41560-017-0025-8

Merk, C., Pönitzsch, G., Rehdanz, K., 2016. Knowledge about aerosol injection does not reduce individual mitigation efforts. *Environmental Research Letters* 11, 054009. https://doi.org/10.1088/1748-9326/11/5/054009

Mersereau, D., 2016. 15 facts about "the year without a summer." *Mental Floss*.

Meyer, W., 2010. The man who saved the whales. *Forbes*. https://www.forbes.com/sites/warrenmeyer/2010/11/05/the-man-who-saved-the-whales/

Millard-Ball, A., 2012. The Tuvalu Syndrome: Can geoengineering solve climate's collective action problem? *Climatic Change* 110, 1047–1066. https://doi.org/10.1007/s10584-011-0102-0

Miller, L., Reynolds, J., 2009. Autism and vaccination – the current evidence. *Journal for Specialists in Pediatric Nursing* 14, 166–172. https://doi.org/10.1111/j.1744-6155.2009.00194.x

Molena, F., 1912. *Popular Mechanics*. 339–342.

Moreno-Cruz, J., Wagner, G., Keith, D., 2018. An

economic anatomy of optimal climate policy (No. ID 3001221), *HKS Faculty Research Working Paper* Series RWP17-028.

Moreno-Cruz, J.B., 2015. Mitigation and the geoengineering threat. *Resource and Energy Economics* 41, 248–263. https://doi.org/10.1016/j.reseneeco.2015.06.001

Morris, E., 2007. From horse power to horsepower. *Access Magazine* 1, 2–10.

Morrow, D.R., 2014. Ethical aspects of the mitigation obstruction argument against climate engineering research. *Philosophical Transactions of the Royal Society A: Mathematical, Physical and Engineering Sciences* 372, 20140062. https://doi.org/10.1098/rsta.2014.0062

Morton, O., 2015. *The Planet Remade: How Geoengineering Could Change the World*. Princeton University Press.

Mufson, S., Mooney, C., Eilperin, J., Muyskens, J., 2019. Extreme climate change has arrived in America. *Washington Post*.

Muri, H., Kristjánsson, J.E., Storelvmo, T., Pfeffer, M.A., 2014. The climatic effects of modifying cirrus clouds in a climate engineering framework. *Journal of Geophysical Research: Atmospheres* 119, 4174–4191. https://doi.org/10.1002/2013JD021063

Nash, J., 1951. Non-cooperative games. *Annals of Mathematics* 54, 286–295. https://doi.org/10.2307/1969529

National Academies of Sciences, Engineering, and Medicine, 2021. *Reflecting Sunlight: Recommendations for Solar Geoengineering and Research Governance*. National Academies Press.

National Research Council, 2015a. *Climate Intervention: Reflecting Sunlight to Cool Earth*. National Academies Press.

National Research Council, 2015b. *Climate Intervention: Carbon Dioxide Removal and Reliable Sequestration*. National Academies Press, Washington, D.C.

National Research Council, 1992. Policy implications of greenhouse warming. National Academy of Sciences Committee on Science, Engineering, and Public Policy, Washington, D.C.

Navarro, J.C.A., Varma, V., Riipinen, I., Seland, Ø., Kirkevåg, A., Struthers, H., Iversen, T., Hansson, H.-C., Ekman, A.M.L., 2016. Amplification of Arctic warming by past air pollution reductions in Europe. *Nature Geosci* 9, 277–281. https://doi.org/10.1038/ngeo2673

Necheles, E., Burns, E.T., Keith, D.W., 2018. Funding for solar geoengineering from 2008 to 2018. Solar Geoengineering Research Blog. https://geoengineering.environment.harvard.edu/blog/funding-solar-geoengineering

Nordhaus, W.D., 2015. Climate clubs: Overcoming free-riding in international climate policy. *American Economic Review* 105, 1339–1370. https://doi.org/10.1257/aer.15000001

Ocko, I.B., Hamburg, S.P., Jacob, D.J., Keith, D.W., Keohane, N.O., Oppenheimer, M., Roy-Mayhew, J.D., Schrag, D.P., Pacala, S.W., 2017. Unmask temporal trade-offs in climate policy debates. *Science* 356, 492–493.

Oman, L., Robock, A., Stenchikov, G., Schmidt, G.A., Ruedy, R., 2005. Climatic response to high-latitude volcanic eruptions. *Journal of Geophysical Research: Atmospheres* 110. https://doi.org/10.1029/2004JD005487

Oreskes, N., 2019. *Why Trust Science?* Princeton University Press.

Parker, A., Irvine, P.J., 2018. The risk of termination

shock from solar geoengineering. *Earth's Future* 6, 456–467.
Parson, E.A., 2003. *Protecting the ozone layer: science and strategy*. Oxford University Press.
Parson, E.A., Ernst, L.N., 2013. International governance of climate engineering. *Theoretical Inquiries in Law* 14, 307–338. https://doi.org/10.1515/til-2013-015
Parson, E.A., Keith, D.W., 2013. End the deadlock on governance of geoengineering research. *Science* 339, 1278–1279.
Pasztor, J., Scharf, C., Schmidt, K.-U., 2017. How to govern geoengineering? *Science* 357, 231–231. https://doi.org/10.1126/science.aan6794
Penna, A.N., Rivers, J.S., 2013. *Natural Disasters in a Global Environment*. John Wiley & Sons.
Peper, E., 2020. *Veil*. Unmapped Press.
Pierrehumbert, R.T., 2019a. There is no Plan B for dealing with the climate crisis. *Bulletin of the Atomic Scientists* 75, 215–221. https://doi.org/10.1080/00963402.2019.1654255
Pierrehumbert, R.T., 2019b. Solar Geoengineering Research Seminar April 24, 2019.
Pierrehumbert, R.T., 2017. The trouble with geoengineers "hacking the planet." *Bulletin of the Atomic Scientists*. https://thebulletin.org/2017/06/the-trouble-with-geoengineers-hacking-the-planet/
Pierrehumbert, R.T., 2015. Climate hacking is barking mad. *Slate Magazine*.
Pigou, A.C., 1920. *The Economics of Welfare*. Palgrave Macmillan.
Pinto, I., Jack, C., Lennard, C., Tilmes, S., Odoulami, R.C., 2020. Africa's climate response to solar radiation management with stratospheric aerosol. *Geophysical Research Letters* 47, e2019GL086047. https://doi.org/10.1029/2019GL086047
Podsakoff, P.M., MacKenzie, S.B., Lee, J.-Y., Podsakoff,

N.P., 2003. Common method biases in behavioral research: A critical review of the literature and recommended remedies. *Journal of Applied Psychology* 88, 879.

Pongratz, J., Lobell, D.B., Cao, L., Caldeira, K., 2012. Crop yields in a geoengineered climate. *Nature Climate Change* 2, 101–105. https://doi.org/10.1038/nclimate1373

Popular Science, 1943. *Why Planes Make Vapor Trails*. Bonnier Corporation.

Poundstone, W., 1993. *Prisoner's Dilemma*. Anchor Books.

Proctor, J., Hsiang, S., Burney, J., Burke, M., Schlenker, W., 2018. Estimating global agricultural effects of geoengineering using volcanic eruptions. *Nature* 560, 480–483. https://doi.org/10.1038/s41586-018-0417-3

Rahman, A.A., Artaxo, P., Asrat, A., Parker, A., 2018. Developing countries must lead on solar geoengineering research. *Nature* 556, 22–24. https://doi.org/10.1038/d41586-018-03917-8

Revelle, R., Broecker, W., Craig, H., Keeling, C.D., Smagorinsky, J., 1965. Atmospheric carbon dioxide, in *Restoring the Quality of Our Environment*. The White House, Washington, D.C.

Reynolds, J., 2015. A critical examination of the climate engineering moral hazard and risk compensation concern. *The Anthropocene Review* 2, 174–191. https://doi.org/10.1177/2053019614554304

Reynolds, J.L., 2019a. *The Governance of Solar Geoengineering: Managing Climate Change in the Anthropocene*. Cambridge University Press, Cambridge, U.K.

Reynolds, J.L., 2019b. Uncovering the origins of false claims in the solar geoengineering discourse. https://geoengineering.environment.harvard.

edu/blog/uncovering-origins-false-claims-solar-geoengineering-discourse
Reynolds, J.L., Parker, A., Irvine, P., 2016. Five solar geoengineering tropes that have outstayed their welcome. *Earth's Future* 4, 562–568.
Reynolds, J.L., Wagner, G., 2019. Highly decentralized solar geoengineering. *Environmental Politics* 1–17. https://doi.org/10.1080/09644016.2019.1648169
Ricke, K. 2019. Solar geo reduces atmospheric CO2 bc it increases uptake of CO2 by land *and ocean* [WWW Document]. Twitter. https://twitter.com/katericke/status/1162076300841902086
Ricke, K., Drouet, L., Caldeira, K., Tavoni, M., 2018. Country-level social cost of carbon. *Nature Climate Change* 1. https://doi.org/10.1038/s41558-018-0282-y
Ridley, M., 2015. Fossil fuels will save the world (really). *Wall Street Journal.*
Robinson, K.S., 2020. *The Ministry for the Future.* Little, Brown Book Group.
Robock, A., 2020. Benefits and risks of stratospheric solar radiation management for climate intervention (geoengineering). *The Bridge* 50, 59–67.
Robock, A., 2015. The CIA asked me about controlling the climate – this is why we should worry. *Guardian.*
Robock, A., 2008. 20 reasons why geoengineering may be a bad idea. *Bulletin of the Atomic Scientists* 64, 14–18. https://doi.org/10.2968/064002006
Robock, A., 2000. Volcanic eruptions and climate. *Reviews of Geophysics* 38, 191–219. https://doi.org/10.1029/1998RG000054
Robock, A., Oman, L., Stenchikov, G.L., 2008. Regional climate responses to geoengineering with tropical and Arctic SO2 injections. *Journal of Geophysical Research: Atmospheres* 113.
Rogelj, J., Meinshausen, M., Knutti, R., 2012. Global

warming under old and new scenarios using IPCC climate sensitivity range estimates. *Nature Climate Change* 2, 248–253. https://doi.org/10.1038/nclimate1385

Roth, A.E., 1993. The early history of experimental economics. *Journal of the History of Economic Thought* 15, 184–209. https://doi.org/10.1017/S1053837200000936

Sandler, T., 2018. Collective action and geoengineering. *Review of International Organizations* 13, 105–125.

Schelling, T.C., 1996. The economic diplomacy of geoengineering. *Climatic Change* 33, 303–307.

Selten, R., 1965. Spieltheoretische behandlung eines Oligopolmodells mit Nachfrageträgheit: Teil i: Bestimmung des dynamischen Preisgleichgewichts. *Zeitschrift für die gesamte Staatswissenschaft/Journal of Institutional and Theoretical Economics* 301–324.

Shabecoff, P., 1988. Global warming has begun, expert tells senate. *New York Times*.

Shearer, C., West, M., Caldeira, K., Davis, S.J., 2016. Quantifying expert consensus against the existence of a secret, large-scale atmospheric spraying program. *Environmental Research Letters* 11, 084011. https://doi.org/10.1088/1748-9326/11/8/084011

Shepherd, M., 2018. There is no NASA "cloud machine" – here's the real explanation of that viral video. *Forbes*.

Sherwood, S., Webb, M.J., Annan, J.D., Armour, K.C., Forster, P.M., Hargreaves, J.C., Hegerl, G., Klein, S.A., Marvel, K.D., Rohling, E.J., Watanabe, M., Andrews, T., Braconnot, P., Bretherton, C.S., Foster, G.L., Hausfather, Z., Heydt, A.S. von der, Knutti, R., Mauritsen, T., Norris, J.R., Proistosescu, C., Rugenstein, M., Schmidt, G.A., Tokarska, K.B., Zelinka, M.D., 2020. An assessment of Earth's climate sensitivity using multiple lines of evidence.

Reviews of Geophysics 58, e2019RG000678. https://doi.org/10.1029/2019RG000678
Shkolnik, A., Taylor, C.R., Finch, V., Borut, A., 1980. Why do Bedouins wear black robes in hot deserts? *Nature* 283, 373–375. https://doi.org/10.1038/283373a0
Smith, W., 2020. The cost of stratospheric aerosol injection through 2100 *Environmental Research Letters*. https://doi.org/10.1088/1748-9326/aba7e7
Smith, W., Wagner, G., 2018. Stratospheric aerosol injection tactics and costs in the first 15 years of deployment. *Environmental Research Letters* 13, 124001. https://doi.org/10.1088/1748-9326/aae98d
Starbuck, A., 1878. *History of the American Whale Fishery from its Earliest Inception to the Year 1876*. The author, Waltham, Massachusetts.
Stockholm International Peace Research Institute, 2018. SIPRI Military Expenditure Database [WWW Document]. https://www.sipri.org/databases/milex
Storelvmo, T., Kristjansson, J.E., Muri, H., Pfeffer, M., Barahona, D., Nenes, A., 2013. Cirrus cloud seeding has potential to cool climate. *Geophysical Research Letters* 40, 178–182. https://doi.org/10.1029/2012GL054201
Stott, P.A., Allen, M., Christidis, N., Dole, R.M., Hoerling, M., Huntingford, C., Pall, P., Perlwitz, J., Stone, D., 2013. Attribution of weather and climate-related events, in Asrar, G.R., Hurrell, J.W. (eds), *Climate Science for Serving Society*. Springer, pp. 307–337.
Stott, P.A., Stone, D.A., Allen, M.R., 2004. Human contribution to the European heatwave of 2003. *Nature* 432, 610–614. https://doi.org/10.1038/nature03089
Szerszynski, B., Kearnes, M., Macnaghten, P., Owen, R., Stilgoe, J., 2013. Why solar radiation management geoengineering and democracy won't mix. *Environment and Planning A* 45, 2809–2816.

Teller, E., 1968. *The Constructive Uses of Nuclear Explosives*. McGraw-Hill, New York, N.Y.

Teller, E., Caldeira, K., Canavan, G., Govindasamy, B., Grossman, A., Hyde, R., Ishikawa, M., Ledebuhr, A., Leith, C., Molenkamp, C., 1999. Long-range weather prediction and prevention of climate catastrophes: A status report (Status Report No. UCRL-JC-135414). Lawrence Livermore National Laboratory, Livermore, California.

Teller, E., Hyde, R., Wood, L., 2002. Active climate stabilization: Practical physics-based approaches to prevention of climate change (No. UCRL-JC-148012). Lawrence Livermore National Laboratory, Livermore, California.

Teller, E., Wood, L., Hyde, R., 1997. Global warming and ice ages: I. Prospects for physics-based modulation of global change, in *The Carbon Dioxide Dilemma*. Lawrence Livermore National Laboratory, Livermore, California, p. 22.

Temple, J., 2019. The U.S. government has approved funds for geoengineering research. *MIT Technology Review.*

Thomson, J.J., 1985. The trolley problem. *Yale Law Journal* 94, 1395–1415.

Tilmes, S., Richter, J.H., Kravitz, B., MacMartin, D.G., Mills, M.J., Simpson, I.R., Glanville, A.S., Fasullo, J.T., Phillips, A.S., Lamarque, J.-F., Tribbia, J., Edwards, J., Mickelson, S., Ghosh, S., 2018. CESM1(WACCM) Stratospheric aerosol geoengineering large ensemble project. *Bulletin of the American Meteorological Society* 99, 2361–2371. https://doi.org/10.1175/BAMS-D-17-0267.1

Tingley, D., Wagner, G., 2017. Solar geoengineering and the chemtrails conspiracy on social media. *Palgrave Communications* 3, 12. https://doi.org/10.1057/s41599-017-0014-3

Trenberth, K.E., Dai, A., 2007. Effects of Mount Pinatubo volcanic eruption on the hydrological cycle as an analog of geoengineering. *Geophysical Research Letters* 34.

Trenberth, K.E., Fasullo, J.T., Shepherd, T.G., 2015. Attribution of climate extreme events. *Nature Climate Change* 5, 725–730. https://doi.org/10.1038/nclimate2657

U.K. Royal Society, 2009. *Geoengineering the Climate: Science, Governance and Uncertainty*. Royal Society, London, U.K.

Urpelainen, J., 2012. Geoengineering and global warming: A strategic perspective. *International Environmental Agreements: Politics, Law and Economics* 12, 375–389. https://doi.org/10.1007/s10784-012-9167-0

U.S. Global Change Research Program, 2019. FY2017–2019 USGCRP Budget Crosscut by [WWW Document]. https://www.globalchange.gov/about/budget

Victor, D.G., 2008. On the regulation of geoengineering. *Oxford Review of Economic Policy* 24, 322–336. https://doi.org/10.1093/oxrep/grn018

Visioni, D., MacMartin, D.G., Kravitz, B., 2021. Is turning down the sun a good proxy for stratospheric sulfate geoengineering? *Journal of Geophysical Research: Atmospheres* n/a, e2020JD033952. https://doi.org/10.1029/2020JD033952

Wagner, G., 2020a. A guide to spending Bezos's new climate war chest. *Bloomberg Green*.

Wagner, G., 2020b. China's carbon neutrality goal is good policy and good politics. *Bloomberg Green* Risky Climate column.

Wagner, G., 2020c. Pausing the world to fight coronavirus has carbon emissions down – but true climate success looks like more action, not less. *TIME*.

Wagner, G., 2020d. Compound growth could kill us – or make us stronger. *Project Syndicate*.

Wagner, G., 2020e. The leadership failure that will cost us everything. *Project Syndicate*.

Wagner, G., 2020f. Carbon taxes alone aren't good climate policy. *Bloomberg Green*.

Wagner, G., 2020g. The true price of carbon. *Project Syndicate*.

Wagner, G., 2018. Chemtrails aren't the geoengineering debate we should be having (because they aren't real). *Earther*. https://earther.gizmodo.com/chemtrails-are-not-the-geoengineering-debate-we-should-1825171856

Wagner, G., 2012. Naomi Klein is half right: Distorted markets are the real problem. *Grist*.

Wagner, G., 2011a. *But Will the Planet Notice?: How Smart Economics Can Save the World*. Hill & Wang/Farrar Strauss & Giroux.

Wagner, G., 2011b. Going green but getting nowhere. *New York Times*.

Wagner, G., Kåberger, T., Olai, S., Oppenheimer, M., Rittenhouse, K., Sterner, T., 2015. Energy policy: Push renewables to spur carbon pricing. *Nature News* 525, 27. https://doi.org/10.1038/525027a

Wagner, G., Merk, C., 2019. Moral hazard and solar geoengineering, in Robert N. Stavins, Robert C. Stowe (eds), *Governance of the Deployment of Solar Geoengineering. Harvard Project on Climate Agreements*. Cambridge, MA, pp. 135–139.

Wagner, G., Merk, C., 2018. The hazard of environmental morality. *Foreign Policy*.

Wagner, G., Samaras, C., 2019. Do we really have only 12 years to avoid climate disaster? *New York Times*.

Wagner, G., Weitzman, M.L., 2018. A big-sky plan to cool the planet. *Wall Street Journal*.

Wagner, G., Weitzman, M.L., 2015. *Climate Shock:*

The Economic Consequences of a Hotter Planet. Princeton University Press.
Wagner, G., Weitzman, M.L., 2012. Playing God. *Foreign Policy.*
Wagner, G., Zizzamia, D., 2021. Green moral hazards. *Ethics, Policy and Environment.* http://dx.doi.org/10.1080/21550085.2021.1940449
Wagner, G., Zizzamia, D., 2020. Green moral hazards. C2G. https://www.c2g2.net/green-moral-hazards/
Watts, J., 2018. We have 12 years to limit climate change catastrophe, warns UN. *Guardian.*
Weaver, R.K., 1986. The politics of blame avoidance. *Journal of Public Policy* 6, 371–398.
Weitzman, M.L., 2015. A voting architecture for the governance of free-driver externalities, with application to geoengineering. *Scandinavian Journal of Economics* 117, 1049–1068. https://doi.org/10.1111/sjoe.12120
Weitzman, M.L., 1974. Prices vs. quantities. *Review of Economic Studies* 41, 477–491.
Wood, G.D., 2014. The volcano that changed the course of history. *Slate Magazine.*
World Health Organization, 2020. Air pollution [WWW Document]. https://www.who.int/westernpacific/health-topics/air-pollution
Xu, C., Kohler, T.A., Lenton, T.M., Svenning, J.-C., Scheffer, M., 2020. Future of the human climate niche. *Proceedings of the National Academy of Sciences.*
Yago, G., 1984. *The Decline of Transit: Urban Transportation in German and U.S. Cities, 1900–1970.* Cambridge University Press.
Yang, J., Peltier, W.R., Hu, Y., 2012. The initiation of modern soft and hard Snowball Earth climates in CCSM4. *Climate of the Past* 8, 907–918. https://doi.org/10.5194/cp-8-907-2012

Yaroshevsky, A.A., 2006. Abundances of chemical elements in the Earth's crust. *Geochemistry International* 44, 48–55. https://doi.org/10.1134/S001670290601006X

Zeckhauser, R.J., 2006. Investing in the unknown and unknowable. *Capitalism and Society* 1/2.

Zeckhauser, R.J., Wagner, G., 2019. The implications of uncertainty and ignorance for solar geoengineering, in Stavins, R.N., Stowe, R.C. (eds), *Governance of the Deployment of Solar Geoengineering*. Cambridge, MA, pp. 107–111.

Zerefos, C.S., Gerogiannis, V.T., Balis, D., Zerefos, S.C., Kazantzidis, A., 2007. Atmospheric effects of volcanic eruptions as seen by famous artists and depicted in their paintings. *Atmospheric Chemistry and Physics* 7, 4027–4042.

Index